Synthesis Lectures on Communications

Series Editor

William H. Tranter, Virginia Tech, Blacksburg, VA, USA

This series of short books cover a wide array of topics, current issues, and advances in key areas of wireless, optical, and wired communications. The series also focuses on fundamentals and tutorial surveys to enhance an understanding of communication theory and applications for engineers.

Jerry D. Gibson

Fourier Transforms, Filtering, Probability and Random Processes

Introduction to Communication Systems

 Springer

Jerry D. Gibson
Department of Electrical and Computer
Engineering
University of California
Santa Barbara, CA, USA

ISSN 1932-1244 ISSN 1932-1708 (electronic)
Synthesis Lectures on Communications
ISBN 978-3-031-19582-2 ISBN 978-3-031-19580-8 (eBook)
https://doi.org/10.1007/978-3-031-19580-8

This Springer imprint is published by the registered company Springer Nature Switzerland AG
The registered company address is: Gewerbestrasse 11, 6330 Cham, Switzerland

Contents

About the Author

Jerry D. Gibson is Professor of Electrical and Computer Engineering at the University of California, Santa Barbara. He is Co-author of the books *Digital Compression for Multimedia* (Morgan-Kaufmann, 1998) and *Introduction to Nonparametric Detection with Applications* (Academic Press, 1975 and IEEE Press, 1995) and Author of the textbook, *Principles of Digital and Analog Communications* (Prentice-Hall, second ed., 1993). He is Editor-in-Chief of *The Mobile Communications Handbook* (CRC Press, 3rd ed., 2012), Editor-in-Chief of *The Communications Handbook* (CRC Press, 2nd ed., 2002), and Editor of the book, *Multimedia Communications: Directions and Innovations* (Academic Press, 2000). His most recent books are *Rate Distortion Bounds for Voice and Video* (Co-author with Jing Hu, NOW Publishers, 2014), and *Information Theory and Rate Distortion Theory for Communications and Compression* (Morgan-Claypool, 2014).

He was Associate Editor for Speech Processing for the *IEEE Transactions on Communications* from 1981 to 1985 and Associate Editor for Communications for the *IEEE Transactions on Information Theory* from 1988–1991. He was an IEEE Communications Society Distinguished Lecturer for 2007–2008.

In 1990, he received The Fredrick Emmons Terman Award from the American Society for Engineering Education, and in 1992, he was elected Fellow of the IEEE. He was the recipient of the 1993 IEEE Signal Processing Society Senior Paper Award for the Speech Processing area. He received the *IEEE Transactions on Multimedia* Best Paper Award in 2010 and the IEEE Technical Committee on Wireless Communications Recognition Award for contributions in the area of wireless communications systems and networks in 2009.

Orthogonal Functions and Fourier Series

1

1.1 Introduction

Fundamental to the practice of electrical engineering and the study of communication systems is the ability to represent symbolically the signals and waveforms that we work with daily. The representation of these signals and waveforms in the time domain is the subject of this chapter. On first thought, writing an expression for some time function seems trivial; that is, everyone can write equations for straight lines, parabolas, sinusoids, and exponentials. Indeed, this knowledge is important, and it is precisely this ability to write expressions for and to manipulate familiar functions that we will draw on heavily in the sequel. Here, however, we are interested in the extension of this ability to include more complex, real-world waveforms, such as those that might be observed using an oscilloscope at the input or output of a circuit or system.

It is intuitive that these more complex signals cannot be fully described by a single member of our set of familiar mathematical functions. Further, often a situation occurs where a signal may not have a unique representation; that is, there may be several different combinations of functions that accurately represent the signal. In this case, however, it usually turns out that one or more of the representations has some advantage over the others, such as fewer terms, ease of manipulation, or being physically more meaningful. Of course, it is not always clear at the outset whether a certain combination of functions has any or all of the advantages above. Of even greater concern is which functions to select and how we are to combine them to obtain a given wave shape without resorting to the cut-and-try method. It is clear that some guidelines are needed here.

We begin our development of these guidelines in Sect. 1.2 by establishing some properties that are desirable for the basic building block functions to have and by introducing particular functions that have these properties. How some of these functions are combined

© The Author(s), under exclusive license to Springer Nature Switzerland AG 2023
J. D. Gibson, *Fourier Transforms, Filtering, Probability and Random Processes*,
Synthesis Lectures on Communications, https://doi.org/10.1007/978-3-031-19580-8_1

to obtain a given waveform and two specific forms of these representations—trigono-metric and complex exponential Fourier series—are developed in Sects. 1.3 and 1.4, respectively. In Sect. 1.5 several properties of waveforms that can greatly simplify the evaluation of Fourier series coefficients are presented and illustrated. Since in many problems an approximation to a waveform is adequate, in Sect. 1.6 we demonstrate that a truncated Fourier series can be used to approximate a function in the least squares sense. Starting with the idea of a least squares approximation, additional details concerning general Fourier series are also given in Sect. 1.6, which provide a much firmer mathematical basis for the Fourier series developments in the preceding sections. In Sect. 1.7 we introduce the important concept of spectral or frequency content of a signal that is used throughout the book and is basic to most facets of electrical engineering.

1.2   Signal Representation and Orthogonal Functions

The ultimate goal of this chapter is to provide the reader with the ability to represent "nonstandard" wave shapes symbolically. One powerful way to approach this problem is to specify a set of basic functions that are then combined in some way to produce expressions for less familiar waveforms. What properties are desirable when selecting this set of basic functions? The answer is available from our knowledge of two- and three-dimensional vector spaces. The fundamental building blocks of these geometrical spaces are unit vectors in the x, y, and z directions, which by definition are *orthonormal,* that is, orthogonal with magnitudes normalized to 1. By analogy, then, it would seem useful to require that our basic building block functions be orthogonal, or even better, orthonormal.

What is the definition of orthogonality in terms of functions? Two real functions $f(t)$ and $g(t)$ are said to be *orthogonal* over the interval (t_0, t_1) if the integral (called the inner product)

$$\int_{t_0}^{t_1} f(t)g(t)\, dt = 0 \tag{1.2.1}$$

for $f(t) \neq g(t)$. Let us test the set of functions $\{1, t, t^2, t^3, \ldots, t^n, \ldots\}$ to see if they are orthogonal. With $f(t) = 1$ and $g(t) = t$, we have

$$\int_{t_0}^{t_1} (1)\, t\, dt = \left.\frac{t^2}{2}\right|_{t_0}^{t_1} = \frac{1}{2}\left[t_1^2 - t_0^2\right].$$

These functions will be orthogonal over symmetrical limits, that is, with $t_0 = -t_1$ Continuing the investigation with $f(t) = 1$ and $g(t) = t^2$, we find that

$$\int_{t_0}^{t_1} (1) t^2 dt = \frac{t^3}{3}\bigg|_{t_0}^{t_1} = \frac{1}{3}[t_1^3 - t_0^3].$$

The functions 1 and t^2 are not orthogonal over a symmetrical interval, and indeed, they do not seem to be orthogonal over any interval, excluding the trivial case when $t_0 = t_1$. This is the general result. The set of functions consisting of powers of t is not orthogonal and hence does not seem to be a good set for representing general wave shapes.

Before continuing the search for orthogonal functions, we extend the definition of orthogonality in Eq. (1.2.1) to complex signals by noting that two complex signals $f(t)$ and $g(t)$ are said to be orthogonal if

$$\int_{t_0}^{t_1} f(t) g^*(t)\, dt = \int_{t_0}^{t_1} f^*(t) g(t)\, dt = 0, \tag{1.2.2}$$

where the superscript * indicates the complex conjugate operation. Furthermore, two possibly complex time functions $f(t)$ and $g(t)$ are said to be ortho*normal* if Eq. (1.2.2) holds and they satisfy the additional relations

$$\int_{t_0}^{t_1} f(t) f^*(t) dt = 1 \quad \text{and} \quad \int_{t_0}^{t_1} g(t) g^*(t) dt = 1. \tag{1.2.3}$$

Notice that Eqs. (1.2.3) are valid for $f(t)$ and $g(t)$ real or complex, since for $f(t)$ and $g(t)$ real we have $f^*(t) = f(t)$ and $g^*(t) = g(t)$.

Now that definitions of orthogonality for both real and complex functions are available, the task remains to find some familiar functions that possess this property. A very important group of functions that are orthogonal is the set of functions $\cos n\omega_0 t$ and $\sin m\omega_0 t$ over the interval $t_0 \le t \le t_0 + 2\pi/\omega_0$ for n and m nonzero integers with $n \ne m$. The demonstration of the orthogonality of these functions is given in the following example.

Example 1.2.1 We would like to investigate the orthogonality of the set of functions $\{\cos n\omega_0 t, \sin m\omega_0 t\}$ over the interval $t_0 \le t \le t_0 + 2\pi/\omega_0$ with n and m nonzero integers and $n \ne m$. Since these functions are real, either Eq. (1.2.1) or (1.2.2) is applicable here. To include all possible combinations of functions, it is necessary that we demonstrate orthogonality for three separate cases. We must show that for $n \ne m$:

(1) $\int_{t_0}^{t_0 + 2\pi/\omega_0} \cos n\omega_0 t \cos m\omega_0 t\, dt = 0$

(2) $\int_{t_0}^{t_0 + 2\pi/\omega_0} \sin n\omega_0 t \sin m\omega_0 t\, dt = 0$

(3) $\int_{t_0}^{t_0 + 2\pi/\omega_0} \cos n\omega_0 t \sin m\omega_0 t\, dt = 0.$

For the first case we have

(1)

$$\int_{t_0}^{t_0+2\pi/\omega_0} \cos n\omega_0 t \cos m\omega_0 t\, dt = \frac{1}{2}\int_{t_0}^{t_0+2\pi/\omega_0} \cos(n+m)\omega_0 t\, dt$$

$$+ \frac{1}{2}\int_{t_0}^{t_0+2\pi/\omega_0} \cos(n-m)\omega_0 t\, dt, \qquad (1.2.4)$$

since $n \neq m$, $n+m$ and $n-m$ are nonzero integers, and the functions $\cos(n+m)\omega_0 t$ and $\cos(n-m)\omega_0 t$ have exactly $n+m$ and $n-m$ complete periods, respectively, in the interval $[t_0, t_0 + 2\pi/\omega_0]$. The integrals in Eq. (1.2.4) thus encompass a whole number of periods and hence both will be zero, since the integral of a cosine over any whole number of periods is zero.

Cases (2) and (3) follow by an identical argument, and hence they are left as an exercise. The reader should note that the case (3) result is also true for $m = n$ (see Problem 1.4).

A very useful set of complex orthogonal functions can be surmised from the set of orthogonal sine and cosine functions just discussed. That is, since $e^{j\theta} = \cos\theta + j\sin\theta$, we are led to conjecture from the immediately preceding results that the functions $e^{jn\omega_0 t}$, $e^{jm\omega_0 t}$ for m and n integers and $m \neq n$ are orthogonal over the interval $[t_0, t_0 + 2\pi/\omega_0]$. The orthogonality of these functions is demonstrated in the next example.

Example 1.2.2 To show that $e^{jn\omega_0 t}$ and $e^{jm\omega_0 t}$ are orthogonal over $t_0 \leq t \leq t_0 + 2\pi/\omega_0$ for m and n integers and $m \neq n$, we employ Eq. (1.2.2) with $t_1 = t_0 + 2\pi/\omega_0$. Substituting $f(t) = e^{jn\omega_0 t}$ and $g^*(t) = e^{-jm\omega_0 t}$ into the integral on the left-hand side of Eq. (1.2.2) produces.

$$\int_{t_0}^{t_0+2\pi/\omega_0} e^{jn\omega_0 t} e^{-jm\omega_0 t}\, dt = \int_{t_0}^{t_0+2\pi/\omega_0} e^{j(n-m)\omega_0 t}\, dt$$

$$= \frac{e^{j(n-m)\omega_0 t_0}}{j(n-m)\omega_0}\left\{ e^{j(n-m)(2\pi)} - 1 \right\}.$$

Since $n-m$ is an integer,

$$e^{j(n-m)(2\pi)} = \cos 2\pi(n-m) + j\sin 2\pi(n-m) = 1,$$

so

$$\int_{t_0}^{t_0+2\pi/\omega_0} e^{jn\omega_0 t} e^{-jm\omega_0 t}\, dt = 0,$$

and hence the set of functions $\{e^{jn\omega_0 t}\}$ for all integral values of n are orthogonal over the interval $[t_0, t_0 + 2\pi/\omega_0]$.

The two examples in this section have demonstrated the orthogonality of two sets of functions that are very important in the analysis and design of communication systems. Some common sets of functions that are not critical to our development but which are orthogonal in the sense that they are orthogonal with respect to a weighting function are Bessel functions, Legendre polynomials, Jacobi polynomials, Laguerre polynomials, and Hermite polynomials. The concept of orthogonality with respect to a weighting function is not required in the sequel and therefore is not considered further here. For additional details, see [1, 2], and Problem 1.5.

Sets of functions that possess the important property of orthogonality were defined and investigated briefly in this section. The task remains for us to demonstrate how the functions in each set can be combined to represent signals and waveforms that occur in communication systems. This development is the subject of the following sections.

1.3   Trigonometric Fourier Series

In the present section we investigate the representation of a waveform in terms of sine and cosine functions. Initially limiting consideration to periodic functions, that is, functions with $g(t) = g(t + T)$ where $T = $ period, we express the periodic function $f(t)$ with period $T = 2\pi/\omega_0$ in terms of an infinite trigonometric series given by

$$f(t) = a_0 + \sum_{n=1}^{\infty} \{a_n \cos n\omega_0 t + b_n \sin n\omega_0 t\}, \tag{1.3.1}$$

where a_0, a_n, and b_n are constants. Notice two things about Eq. (1.3.1). First, the reason we are working only with periodic functions is that each of the terms on the right-hand side of Eq. (1.3.1) is periodic. Second, an infinite number of terms are included in the representation. Why this is necessary is discussed later in the section and in more detail in Sect. 1.6.

Before the trigonometric series in Eq. (1.3.1) can be called a Fourier series, it remains to specify the constant coefficients a_0, a_n, and b_n, $n = 1, 2, \ldots$ These coefficients can be determined as follows. To find a_0, multiply both sides of Eq. (1.3.1) by dt and integrate over one (arbitrary) period $t_0 \le t \le t_0 + T$, to obtain

$$a_0 = \frac{1}{T} \int_{t_0}^{t_0+T} f(t) dt. \tag{1.3.2}$$

To derive an expression for a_n, we multiply both sides of Eq. (1.3.1) by $\cos k\omega_0 t \, dt$ and integrate over a period, which produces (letting $k \to n$)

$$a_n = \frac{2}{T} \int_{t_0}^{t_0+T} f(t) \cos n\omega_0 t \, dt \qquad (1.3.3)$$

for $n = 1, 2, 3, \ldots$ Similarly, an expression for the b_n coefficients is obtained by multiplying Eq. (1.3.1) by $\sin k\omega_0 t \, dt \, dt$ and integrating over one period, which yields

$$b_n = \frac{2}{T} \int_{t_0}^{t_0+T} f(t) \sin n\omega_0 t \, dt. \qquad (1.3.4)$$

Equation (1.3.1) with the coefficients defined by Eqs. (1.3.2), (1.3.3), and (1.3.4) is called a *trigonometric Fourier series.* Any periodic function can be expanded in a Fourier series simply by determining the period T and using these equations. As an illustration of the procedure, consider the following example.

Example 1.3.1 We would like to write a Fourier series representation of the waveform shown in Fig. 1.1. To do this, we first find the period of the waveform, which is $T = \tau$. Next we must pick the period that we wish to integrate over, which is equivalent to selecting t_0 in the expressions for the coefficients. Since this choice is arbitrary, it is wise to select that period which most simplifies the evaluation of the integrals. For this particular case, we choose $t_0 = -\tau/2$.

We are now ready to evaluate the coefficients. Using Eq. (1.3.2) with $T = \tau$ and $t_0 = -\tau/2$, we have

$$a_0 = \frac{1}{\tau} \int_{-\tau/2}^{\tau/2} f(t) dt = \frac{1}{\tau} \int_{-\tau/4}^{\tau/4} (1) dt = \frac{1}{2}. \qquad (1.3.5)$$

Calculating a_n using Eq. (1.3.3), we see that

$$a_n = \frac{2}{\tau} \int_{-\tau/2}^{\tau/2} f(t) \cos n\omega_0 t \, dt$$

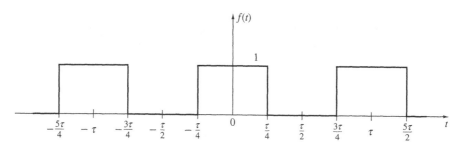

Fig. 1.1 Waveform for Example 1.3.1

$$= \frac{2}{\tau} \cdot \frac{1}{n\omega_0} \sin n\omega_0 t \Bigg|_{-\tau/4}^{\tau/4} = \frac{1}{n\pi} \left[\sin \frac{n\pi}{2} - \sin \frac{-n\pi}{2} \right]. \qquad (1.3.6)$$

Noting that the since function is odd, that is, $\sin(-x) = -\sin x$, Eq. (1.3.6) can be written compactly as

$$a_n = \begin{cases} \frac{2}{n\pi}(-1)^{(n-1)/2}, & \text{for } n \text{ odd} \\ 0, & \text{for } n \text{ even} \end{cases} \qquad (1.3.7)$$

since the sine of integral multiples of π radians is zero and the sine of integral multiples of $\pi/2$ is $+1$.

For the b_n coefficients, we find from Eq. (1.3.4) that

$$b_n = \frac{2}{\tau} \int_{-\tau/2}^{\tau/2} f(t) \sin n\omega_0 t \, dt = \frac{2}{\tau} \int_{-\tau/4}^{\tau/4} (1) \sin n\omega_0 t \, dt = 0. \qquad (1.3.8)$$

By substituting Eqs. (1.3.5), (1.3.7), and (1.3.8) into Eq. (1.3.1), the desired Fourier series representation is

$$f(t) = \frac{1}{2} + \sum_{\substack{n=1 \\ n \text{ odd}}}^{\infty} \frac{2}{n\pi}(-1)^{(n-1)/2} \cos n\omega_0 t, \qquad (1.3.9)$$

where $\omega_0 = 2\pi/\tau$.

It is important to notice that the Fourier series expansion of a periodic function is valid for all time, $-\infty < t < \infty$, even though the integration when computing the coefficients is carried out over only one period. This is because since the function is periodic, if we accurately represent the function over one period, we have an accurate representation for all other periods, and hence for all time.

To obtain a Fourier series representation of a nonperiodic function over a given finite interval, the approach is to let the time interval of interest be the period, T, and proceed exactly as before. That is, the coefficients are evaluated as if the function were periodic. The resulting Fourier series is an exact representation of the function within the time interval, which was assumed to be one period. The Fourier series may be totally inaccurate outside this time interval; however, this is of no consequence to us. The principal difference between writing a Fourier series for a periodic or a nonperiodic waveform is that in the periodic case, the series is an accurate expression for all time, whereas in the nonperiodic case, the series is valid only over the time interval assumed to be one period.

1.4   Exponential (Complex) Fourier Series

The exponential or complex form of a Fourier series is extremely important to our study of communication systems, although just how important this form of Fourier series is will not become clear until Sect. 1.7 and Chap. 2. The complex form can be obtained from the trigonometric Fourier series in Eq. (1.3.1) by some simple manipulations. What is required is to note that sine and cosine functions can be written in terms of complex exponentials as

$$\sin x = \frac{1}{2j}\left[e^{jx} - e^{-jx}\right]$$

and

$$\cos x = \frac{1}{2}\left[e^{jx} + e^{-jx}\right].$$

Substituting these expressions into Eq. (1.3.1) produces

$$f(t) = a_0 + \sum_{n=1}^{\infty}\left\{a_n\left[\frac{1}{2}\left(e^{jn\omega_0 t} + e^{-jn\omega_0 t}\right)\right] + b_n\left[\frac{1}{2j}\left(e^{jn\omega_0 t} - e^{-jn\omega_0 t}\right)\right]\right\}$$

$$= a_0 + \sum_{n=1}^{\infty}\left\{\left(\frac{a_n - jb_n}{2}\right)e^{jn\omega_0 t} + \left(\frac{a_n + jb_n}{2}\right)e^{-jn\omega_0 t}\right\} \qquad (1.4.1)$$

after grouping like exponential terms. If we define the complex Fourier coefficient

$$c_0 = a_0, \quad c_n = \frac{a_n - jb_n}{2}, \quad c_{-n} = \frac{a_n + jb_n}{2} \qquad (1.4.2)$$

for $n = 1, 2, \ldots$, Eq. (1.3.1) can be written as

$$f(t) = \sum_{n=-\infty}^{\infty} c_n e^{jn\omega_0 t} \qquad (1.4.3)$$

which is the exponential or complex form of a Fourier series. Notice in Eq. (1.4.3) that the summation over n extends from $-\infty$ to $+\infty$.

We still need expressions for the complex coefficients, c_n, in terms of $f(t)$, the function to be expanded. The required relationship can be determined from the definitions in Eq. (1.4.2) or by starting with Eq. (1.4.3) and using the orthogonality property of complex exponentials proved in Example 1.2.2. The final result is that

$$c_n = \frac{1}{T}\int_{t_0}^{t_0+T} f(t)e^{-jn\omega_0 t}\,dt, \qquad (1.4.4)$$

where $n = \ldots, -2, -1, 0, 1, 2, \ldots$

Example 1.4.1 To illustrate the calculation of the complex Fourier coefficients, let us obtain the exponential Fourier series representation of the waveform for Example 1.3.1 shown in Fig. 1.1. Since $T = \tau$ and $\omega_0 = 2\pi / \tau$, again let $t_0 = -\tau / 2$, so that Eq. (1.4.4) becomes.

$$c_n = \frac{1}{\tau} \int_{-\tau/4}^{\tau/4} e^{-jn\omega_0 t} dt = \frac{1}{n\pi} \sin \frac{n\pi}{2}. \qquad (1.4.5)$$

The complex Fourier coefficients are thus given by

$$c_n = \begin{cases} 0, & \text{for } n \text{ even} \\ \frac{1}{n\pi}(-1)^{(n-1)/2}, & \text{for } n \text{ odd} \\ \frac{1}{2}, & \text{for } n = 0 \end{cases} \qquad (1.4.6)$$

using L'Hospital's rule (see Problem 1.12) to evaluate c_n for $n = 0$. Actually, it is always preferable to evaluate Eq. (1.4.4) for the $n \neq 0$ and the $n = 0$ cases separately to minimize complications. The exponential Fourier series is given by Eq. (1.4.3), with the coefficients displayed in Eq. (1.4.6).

As in the case of trigonometric Fourier series, a complex Fourier series can be written for a nonperiodic waveform provided that the time interval of interest is assumed to be one period. Just as before, the resulting Fourier series may not be accurate outside the specified time interval of interest. The trigonometric Fourier series in Sect. 1.3 and the complex Fourier series discussed in this section are not actually two different series but are simply two different forms of the same series. There are situations where one form may be preferred over the other, and hence these alternative forms provide us with additional flexibility in signal analysis work if we are familiar with both versions. No other forms of Fourier series are treated in detail in the body of the text, since the trigonometric and complex Fourier series will prove completely satisfactory for our purposes. In the following sections we investigate Fourier series in more depth and study ways to simplify the evaluation of Fourier series coefficients.

1.5 Fourier Coefficient Evaluation Using Special Properties

The calculation of the Fourier series coefficients by straightforward evaluation of the integrals as was done in Sects. 1.3 and 4 can sometimes be quite tedious. Fortunately, in a few special cases, we can obtain relief from part of this computational burden, depending on the properties of the waveform, say $f(t)$, that we are trying to represent.

Probably the most used properties when evaluating trigonometric Fourier series are even and odd symmetry. A waveform has *even symmetry* or is *even* if $f(t) = f(-t)$, and a waveform is *odd* or has *odd symmetry* if $f(t) = -f(-t)$. The trigonometric Fourier series representation of an even waveform contains no sine terms, that is, the $b_n = 0$ for all n, and the representation of an odd function has no constant term or cosine terms, so the $a_n = 0$ for $n = 0, 1, 2, \ldots$ It is straightforward to demonstrate these results by using the even and odd symmetry properties in the expressions for a_n and b, in Eqs. (1.3.3) and (1.3.4).

Example 1.5.1 The waveform in Fig. 1.1 for Example 1.3.1 is even, since $f(t) = f(-t)$ for all values of t. As a result, we should have found in Example 1.3.1 that $b_n = 0$ for all n. This is exactly as we determined, as can be seen from Eq. (1.3.8).

Example 1.5.2 Suppose that we wish to obtain a trigonometric Fourier series representation for the function shown in Fig. 1.2. If possible, we would like to simplify the required calculations by using either the even or the odd symmetry property. Checking to see whether the function is even or odd, we find that $f(t) \neq f(-t)$ but $f(t) = -f(-t)$, and hence the waveform is odd. From our earlier results, then, we conclude that $a_n = 0$ for $n = 0, 1, 2, \ldots$. The calculation of the b_n coefficients is left as an exercise for the reader.

Although it is not necessary for any of the results obtained thus far in the book, we have implicitly assumed throughout that all the waveforms and functions that we are representing by a Fourier series are purely real. This assumption is entirely acceptable since all the waveforms that we can observe in communication systems are real. There are times, however, when it is mathematically convenient to work with complex signals. When we use complex functions in any of our mathematical developments, this fact will be stated explicitly or it will be clear from the context.

Limiting consideration to purely real functions, we can also use even and odd symmetry to simplify or check the calculation of complex Fourier coefficients for waveforms that have either of these properties. Specifically, for $f(t)$ real and even, the complex Fourier coefficients given by Eq. (1.4.4) are all real, while for $f(t)$ real and odd, the c_n are all imaginary. These statements can be proven in a number of ways. The most transparent approach (assuming no knowledge of the a_n and b_n properties) is to use Euler's identity to rewrite Eq. (1.4.4) and then apply the definitions of even and odd functions. Of course, if it is known that $a_n = 0$ for $f(t)$ odd and $b_n = 0$ for $f(t)$ even, the c_n properties can be established by inspection of Eq. (1.4.2).

Example 1.5.3 In Example 1.5.1 the waveform in Fig. 1.1 was shown to be even. Since this function is also real, we conclude that the complex Fourier coefficients should be purely real. Examining the results of Example 1.4.1, we see that the c_n are indeed purely real, just as expected.

Example 1.5.4 The waveform in Fig. 1.2 was concluded to be an odd function in Example 1.5.2. The waveform is also real, and hence the complex Fourier coefficients should be

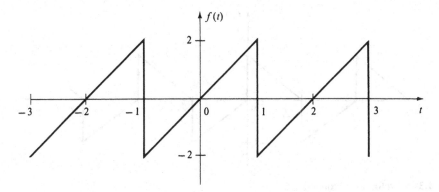

Fig. 1.2 Waveform for Example 1.5.2

purely imaginary. Noting that the period of the waveform is $T = 2$ and hence $\omega_0 = \pi$, we have from Eq. (1.4.4) with $t_0 = 1$,

$$c_n = \frac{1}{2} \int_{-1}^{1} 2te^{-jn\omega_0 t} dt = \frac{j2}{n\pi} \cos n\pi, \tag{1.5.1}$$

which is clearly purely imaginary, as predicted.

Another interesting and useful property is that of *rotation symmetry*, which means that $f(t) = -f(t \pm T/2)$, where T is the period of $f(t)$. A function that has rotation symmetry is also said to be *odd harmonic*, since its complex Fourier coefficients are nonzero only for n odd.

Example 1.5.5 The signal in Fig. 1.3 has the property that $f(t) = -f(t \pm T/2)$, and thus its complex Fourier coefficients should be zero for $n = 0, 2, 4, \ldots$. By direct evaluation from Eq. (1.4.4) for $n \neq 0$,

$$c_n = \frac{1}{T} \int_{-T/2}^{T/2} f(t)e^{-jn\omega_0 t} dt$$

$$= \frac{1}{T} \left\{ \frac{1}{jn\omega_0} - \frac{4}{Tn^2\omega_0^2} \right\} [\cos n\pi - 1]. \tag{1.5.2}$$

When n is even, $\cos n\pi = +1$ and the $c_n = 0$; further, when n is odd, $\cos n\pi = -1$ and the c_n, are nonzero. Finally, for $n = 0$ in Eq. (1.4.4), $c_0 = 0$. This is the desired result.

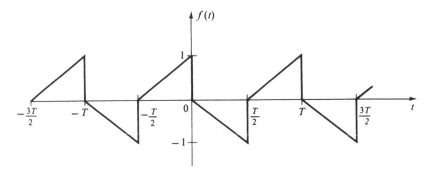

Fig. 1.3 Waveform for Example 1.5.5

As has been demonstrated by examples, the recognition that a signal waveform has one of the special properties mentioned can save substantial amounts of time and effort in the evaluation of Fourier series coefficients. The properties can also provide excellent checks on the accuracy of coefficient calculations. Because of these facts, the reader should become as familiar as possible with the special properties and their application.

1.6   Least Squares Approximations and Generalized Fourier Series

An important question to be answered concerning Fourier series, especially for some practical applications, is: What kind of approximation do we have if we truncate a Fourier series and retain only a finite number of terms? To answer this question, let the truncated trigonometric Fourier series of $f(t)$ be denoted by

$$f_N(t) = a_0 + \sum_{n=1}^{N} \{a_n \cos n\omega_0 t + b_n \sin n\omega_0 t\}. \tag{1.6.1}$$

It is possible to show that the coefficients of the truncated Fourier series in Eq. (1.6.1) are precisely those coefficients that minimize the integral squared error (ISE):

$$\text{ISE} = \int_{t_0}^{t_0+T} [f(t) - h_N(t)]^2 dt, \tag{1.6.2}$$

where

$$h_N(t) = p_0 + \sum_{n=1}^{N} [p_n \cos n\omega_0 t + q_n \sin n\omega_0 t]. \tag{1.6.3}$$

In other words, the partial sum $f_N(t)$, which is generated by truncating the Fourier series expansion of $f(t)$, is *the* one of all possible trigonometric sums $h_N(t)$ of order N or less that minimizes the ISE in Eq. (1.6.2). The sum $f_N(t)$ is usually said to approximate $f(t)$ in the least squares sense.

The proof of this result can be approached in several ways. One way is just direct substitution; another way is to use Eq. (1.6.3) for $h_N(t)$ in Eq. (1.6.2), take partial derivatives with respect to each of the coefficients, equate each derivative to zero, and solve for p_n, and q_n. Note that this approach yields only necessary conditions on the coefficients to minimize Eq. (1.6.2) (see Problem 1.21).

For some applications an approximation to a waveform may prove adequate, and in these cases it is necessary to determine the accuracy of the approximation. A natural indicator of approximation accuracy is the ISE in Eq. (1.6.2) with $h_N(t)$ replaced by the approximation being used. In our case, $h_N(t) = f_N(t)$, which minimizes Eq. (1.6.2). A simplified form of the minimum value of the integral squared error in terms of $f(t)$ and the Fourier coefficients can be shown by straightforward manipulations to be

$$\int_{t_0}^{t_0+T} [f(t) - f_N(t)]^2 dt = \int_{t_0}^{t_0+T} f^2(t)dt - \left\{ Ta_0^2 + \frac{T}{2}\sum_{n=1}^{N}[a_n^2 + b_n^2] \right\}. \quad (1.6.4)$$

Hence for a given signal $f(t)$, Eq. (1.6.4) can be used to calculate the error in the approximation $f_N(t)$, and the number of terms N can be increased until the approximation error is acceptable. Notice that since we have obtained Eq. (1.6.4) by letting $h_N(t) = f_N(t)$ in Eq. (1.6.2), Eq. (1.6.4) is the minimum value of the ISE possible using trigonometric sums.

The following example illustrates the approximation of a given waveform to a prespecified degree of accuracy.

Example 1.6.1 We would like to obtain a truncated trigonometric series approximation to $f(t)$ in Fig. 1.4 such that the ISE in the approximation is 2% or less of the integral squared value of $f(t)$. Note that since $f(t)$ is periodic, it exists for all time, and hence the integral squared value of $f(t)$ is actually infinite. However, if we consider only one period, as was done in deriving Eq. (1.6.4), no difficulties arise.

Referring to Fig. 1.4, we see that with $t_0 = 0$,

$$\int_{t_0}^{t_0+T} f^2(t)dt = \int_{0}^{2\pi} f^2(t)dt = \int_{0}^{\pi} \sin^2 t\, dt = \frac{\pi}{2} = 1.571. \quad (1.6.5)$$

The trigonometric Fourier series coefficients for $f(t)$ are

$$a_0 = \frac{1}{\pi}, \quad a_n = \frac{(-1)^{n+1} - 1}{\pi(n^2 - 1)}, \quad b_1 = \frac{1}{2}, \quad b_n = 0 \quad \text{for } n \neq 1.$$

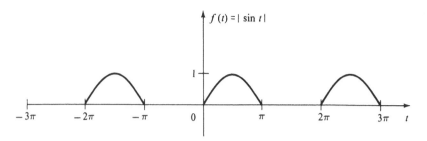

Fig. 1.4 Waveform for Example 1.6.1

First letting $N = 1$ in Eq. (1.6.1), we have $a_0 = 1/\pi$, $a_1 = 0$, $b_1 = \frac{1}{2}$, so Eq. (1.6.4) becomes, upon using Eq. (1.6.5),

$$\int_0^{2\pi} [f(t) - f_1(t)]^2 dt = 1.571 - \left\{ 2\pi \left(\frac{1}{\pi}\right)^2 + \pi \left(\frac{1}{2}\right)^2 \right\} = 0.149. \tag{1.6.6}$$

Dividing by Eq. (1.6.5), we find that the ISE is 9.5% of the integral squared value of $f(t)$, which is above our goal.

Upon letting $N = 2$, the required Fourier series coefficients are $a_0 = 1/\pi$, $a_1 = 0$, $a_2 = -2/3\pi$, $b_1 = \frac{1}{2}$ and $b_n = 0$ for $n \neq 1$. Substituting into Eq. (1.6.4), we find that

$$\int_0^{2\pi} [f(t) - f_2(t)]^2 dt = 1.571 - \left\{ 2\pi \left(\frac{1}{\pi}\right)^2 + \pi \cdot \left[\frac{4}{9\pi^2} + \frac{1}{4}\right] \right\} = 0.0075, \tag{1.6.7}$$

which is only 0.5% of the integral squared value of $f(t)$. The desired approximation is thus

$$f_2(t) = \frac{1}{\pi} + \frac{1}{2} \sin t - \frac{2}{3\pi} \cos 2t. \tag{1.6.8}$$

Thus far in this section, the discussions have been limited to trigonometric Fourier series. Of course, similar results can be obtained for the exponential Fourier series defined in Sect. 1.4. However, rather than consider several specific cases, let us define what is called a generalized Fourier series given by

$$f(t) = \sum_{n=-\infty}^{\infty} \gamma_n \phi_n(t), \tag{1.6.9}$$

where $f(t)$ is some waveform defined over an interval $t_0 \leq t \leq t_0 + T$, the γ_n are the coefficients yet to be determined, and the $\{\phi_n(t)\}$, $n = \ldots, -2, -1, 0, 1, 2\ldots$, are a set of possibly complex orthogonal functions over $t_0 \leq t \leq t_0 + T$. Since the $\phi_n(t)$ are orthogonal, they satisfy the relation

$$\int_{t_0}^{t_0+T} \phi_n(t)\phi_m^*(t)dt = \begin{cases} 0, & \text{for } m \neq n \\ K_n, & \text{for } m = n. \end{cases} \tag{1.6.10}$$

Expressions for the coefficients γ_n, can be obtained by multiplying both sides of Eq. (1.6.9) by $\phi_m^*(t)$ and integrating from t_0 to $t_0 + T$, so that

$$\gamma_n = \frac{1}{K_n} \int_{t_0}^{t_0+T} f(t)\phi_n^*(t)dt \tag{1.6.11}$$

for all n. The derivation of Eq. (1.6.11) is left as an exercise for the reader (see Problem 1.25).

In deriving Eq. (1.6.11) and the equivalent expressions for the trigonometric and complex Fourier series coefficients, it is necessary to integrate the postulated series, Eq. (1.6.9) here, term by term. We have yet to address the validity of this approach. Although we will not prove the result, the only requirement is that the original series formulation, Eqs. (1.3.1), (1.4.3), and (1.6.9), be uniformly convergent for all t in the interval being considered. Hence assuming uniform convergence of the various forms of Fourier series, our derivations of the Fourier coefficients are justified. (See [2] for a discussion of uniform convergence.)

It is possible to show that the truncated generalized Fourier series given by

$$g_N(t) = \sum_{n=-N}^{N} \gamma_n \phi_n(t) \tag{1.6.12}$$

minimizes the integral squared error

$$\varepsilon^2 = \int_{t_0}^{t_0+T} |f(t) - h_N(t)|^2 dt$$

$$= \int_{t_0}^{t_0+T} [f(t) - h_N(t)][f(t) - h_N(t)]^* dt \tag{1.6.13}$$

for all $h_N(t)$ of the form $h_N(t) = \sum_{n=-N}^{N} p_n \phi_n(t)$. Furthermore, the minimum value of the ISE can be shown to be

$$\varepsilon_{\min}^2 = \int_{t_0}^{t_0+T} |f(t)|^2 dt - \sum_{n=-N}^{N} K_n |\gamma_n|^2. \tag{1.6.14}$$

It is interesting to note that using Eq. (1.6.9),

$$\int_{t_0}^{t_0+T} |f(t)|^2 dt = \int_{t_0}^{t_0+T} f(t) \left\{ \sum_{n=-\infty}^{\infty} \gamma_n \phi_n(t) \right\}^* dt$$

$$= \sum_{n=-\infty}^{\infty} \gamma_n^* \int_{t_0}^{t_0+T} f(t)\phi_n^*(t)dt$$

$$= \sum_{n=-\infty}^{\infty} K_n\gamma_n\gamma_n^* = \sum_{n=-\infty}^{\infty} K_n|\gamma_n|^2, \qquad (1.6.15)$$

which is known as *Parseval's theorem*. As a consequence of Eq. (1.6.15), we see that

$$\lim_{N\to\infty} \varepsilon_{\min}^2 = 0, \qquad (1.6.16)$$

which simply states that the generalized Fourier series in Eq. (1.6.9) represents $f(t)$ exactly over the specified interval in the sense of providing a minimum integral squared error. Equation (1.6.15) also indicates that the integral squared value of a signal $f(t)$ can be expressed in terms of the generalized Fourier series coefficients.

Finally, we would like to consider the question: How many terms are necessary to represent a waveform exactly? We have, of course, already answered this question by letting our Fourier series representations have an infinite number of terms. What we would like to do now is to indicate why this was done. Central to the discussion are the concepts of *completeness* and *uniqueness*. Definitions of these terms are available in advanced calculus books [2]. Completeness is concerned with the fact that the $ISE \to 0$ as $N \to \infty$ [see Eq. (1.6.16)], while uniqueness is related to the requirement of having enough functions to represent a given waveform. If a set of orthogonal functions is complete, the set of functions also has the uniqueness property. However, uniqueness alone does not guarantee that a set of functions is complete. What we were doing then, when we included an infinite number of terms in the Fourier series representations in this and earlier sections, was to ensure that the set of orthogonal functions we were using was complete. The trigonometric functions including the constant term in Sect. 1.3 and the complex exponential functions in Sect. 1.4 are all complete, and therefore, Fourier series representations in terms of these functions are unique.

1.7   Spectral Content of Periodic Signals

The trigonometric and complex exponential Fourier series representations in Sects. 1.3 and 1.4 are methods of separating a periodic time function into its various components. In this section we explicitly emphasize the frequency content interpretation of the trigonometric and complex Fourier series coefficients. Although both the trigonometric and complex forms contain identical information, the complex exponential form is used as a basis for the development here. The primary reasons for this selection are that the complex coefficients require the evaluation of only a single integral, and more important, the exponential form leads us rather directly to the definition of the Fourier transform, as will be seen in Chap. 2.

The set of complex Fourier coefficients c_n, completely describe the frequency content of a periodic signal $f(t)$, and as a group constitute what is usually called the *line spectrum* or simply spectrum of $f(t)$. The motivation for the former terminology will be clear soon. Each coefficient specifies the complex amplitude of a certain frequency component. For example, the coefficient c_0 is the amplitude of the zero-frequency value, usually called the *average* or *direct-current* value. The coefficient c_1 indicates the amplitude of the component with the same period as $f(t)$, and hence is called the *fundamental frequency*. The other c_n represent the complex amplitudes of the n harmonics of $f(t)$.

To aid in visualizing the frequency content of a periodic signal, it is possible to sketch several different graphical representations of the complex Fourier coefficients. Since the c_n are in general complex, two graphs are necessary to display all of the information completely. One possible pair of graphs would be to plot the real and imaginary parts of c_n, however, it is more common to plot the magnitude and phase of the c_n given by Eqs. (1.7.1) and (1.7.2), respectively,

$$|c_n| = \left\{ [\text{Re}(c_n)]^2 + [\text{Im}(c_n)]^2 \right\}^{1/2} \tag{1.7.1}$$

and

$$\angle c_n = \tan^{-1} \frac{\text{Im}(c_n)}{\text{Re}(c_n)}. \tag{1.7.2}$$

Equations (1.7.1) and (1.7.2) are usually called the *amplitude spectrum* and *phase spectrum* of $f(t)$, respectively. The magnitude and phase plots are important, since they can be related to the amplitude and phase of a sinusoid (see Problem 1.11). Of course, both pairs of graphs contain the same information. Important properties of the amplitude and phase spectrum are that the amplitude is even,

$$|c_n| = |c_{-n}| \tag{1.7.3}$$

and the phase is odd,

$$\angle c_n = -\angle c_{-n}. \tag{1.7.4}$$

These results provide very tangible benefits, since it is only necessary to determine the magnitude and phase for positive n and then use Eqs. (1.7.3) and (1.7.4) to determine the magnitude and phase for negative n.

Example 1.7.1 We would like to sketch the amplitude and phase spectra of the square wave in Fig. 1.5. Of course, to determine the amplitude and phase spectra, it is first necessary to compute the complex Fourier series coefficients. Using Eq. (1.4.4) directly with $t_0 = 0$ and $T = \tau$, we have for $n \neq 0$,

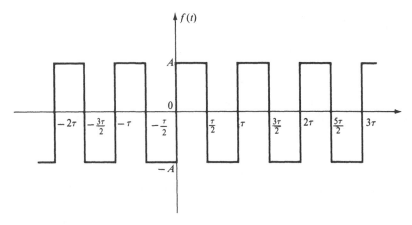

Fig. 1.5 Waveform for Example 1.7.1

$$c_n = \begin{cases} 0, & \text{for } n \text{ even} \\ \frac{-j2A}{n\pi}, & \text{for } n \text{ odd} \end{cases} \qquad (1.7.5)$$

and $c_0 = 0$.

Let us now use some of the special properties developed in Sect. 1.5 and this section to check these results. From Fig. 1.5 we can see immediately that $f(t)$ is real and odd and hence the c_n should be purely imaginary. This checks with Eq. (1.7.5). Observe, too, that $f(t)$ satisfies the rotation symmetry property, that is, $f(t) = -f(t + T/2)$, and hence the c_n should be nonzero only for n odd, which is in agreement with our results. We also know that for $f(t)$ real, $c_{-n} = c_n^*$, which is seen to be true from Eq. (1.7.5) by replacing n with $-n$.

We now calculate the magnitude and phase of the c_n. It is only necessary to find the spectra for $n > 0$, since Eqs. (1.7.3) and (1.7.4) will then give us the results for $n < 0$. Using Eq. (1.7.1), we obtain the amplitude spectrum

$$|c_n| = \{[\text{Re}(c_n)]^2 + [\text{Im}(c_n)]^2\}^{1/2} = \left\{(0)^2 + \left(\frac{-2A}{n\pi}\right)^2\right\}^{1/2} = \frac{2A}{n\pi} = |c_{-n}| \qquad (1.7.6)$$

for n odd and employing Eq. (1.7.3). For n even,

$$|c_n| = |c_{-n}| = 0. \qquad (1.7.7)$$

The phase spectrum for $n > 0$ is given by Eq. (1.7.2) as

Fig. 1.6 Values of the tangent function at 90° intervals

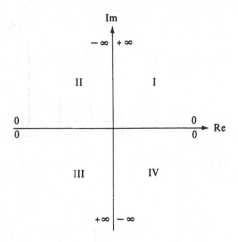

$$c_n = \tan^{-1}\frac{\text{Im}(c_n)}{\text{Re}(c_n)} = \tan^{-1}\left(\frac{-2A/n\pi}{0}\right). \tag{1.7.8}$$

Our first impulse is to conclude from Eq. (1.7.8) that $c_n = \tan^{-1}(-\infty)$. However, a sticky problem suddenly arises since $\tan^{-1}(-\infty)$ is a multivalued function. (See any set of mathematical tables for a plot of the tangent function: e.g., Selby and Girling [1965]).

The problem can be resolved in an unambiguous manner by sketching the complex plane and writing the values of the tangent function at 90° intervals as shown in Fig. 1.6. Upon reconsidering Eq. (1.7.8), the negative imaginary part puts us either in quadrant III or IV. In these two quadrants, the tangent takes on the value of $-\infty$ at only one angle, namely, $-\pi/2$ radians. Hence for $n > 0$ and n odd,

$$\angle c_n = -\frac{\pi}{2}. \tag{1.7.9}$$

We are now ready to plot the amplitude and phase spectra of $f(t)$. What parameter should be used for the abscissa? We could use n, but we can also use $n\omega_0$, since c_n is the complex amplitude of the nth harmonic. Since we desire a frequency spectrum interpretation here, $n\omega_0$ will be used as the abscissa, where $\omega_0 = 2\pi/\tau$. Figures 1.7 and 1.8 show sketches of the amplitude and phase spectra of $f(t)$, respectively, obtained from Eqs. (1.7.6), (1.7.7), and (1.7.9). The most noticeable thing about Figs. 1.7 and 1.8 is that both the amplitude and phase spectra are discrete; that is, they are defined only for discrete values of frequency that are integral multiples of c_0. This is not surprising, though, since any periodic signal has discrete amplitude and phase spectra because only integral multiples of the fundamental frequency are required to synthesize the waveforms via a Fourier series. The terminology *line spectrum* is drawn from the distinctive appearance illustrated by Figs. 1.7 and 1.8. As mentioned earlier, the entire set of complex Fourier coefficients is usually called the spectrum of $f(t)$.

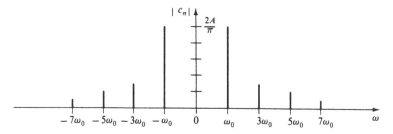

Fig. 1.7 Amplitude spectrum for Example 1.7.1

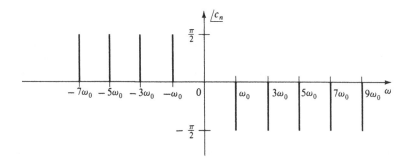

Fig. 1.8 Phase spectrum for Example 1.7.1

Since the last example was somewhat drawn out, let us work an additional example in a very concise fashion.

Example 1.7.2 We desire to sketch the amplitude and phase spectra of $f(t)$ in Fig. 1.1. In Example 1.4.1 the complex Fourier series coefficients for this waveform were found to be.

$$
c_n = \begin{cases} 0, & \text{for } n \text{ even} \\ \frac{1}{n\pi}(-1)^{(n-1)/2}, & \text{for } n \text{ odd} \\ \frac{1}{2}, & \text{for } n = 0. \end{cases}
$$

Since the amplitude spectrum is even and the phase spectrum is odd, it is only necessary to consider $n \geq 0$.

The c_n are purely real for this waveform, so from Eq. (1.7.1) for n odd,

$$
c_n = \left\{ \left[\frac{1}{n\pi}(-1)^{(n-1)/2} \right]^2 \right\}^{1/2} = \frac{1}{n\pi}, \tag{1.7.10}
$$

since $(-1)^{n-1} = +1$ for n odd. The values for $n = 0$ and n even can be obtained by inspection, so in summary

$$|c_n| = \begin{cases} 0, & \text{for } n \text{ even} \\ \frac{1}{n\pi}, & \text{for } n \text{ odd} \\ \frac{1}{2}, & \text{for } n = 0. \end{cases} \tag{1.7.11}$$

For the phase spectrum with n odd, we have from Eq. (1.7.2),

$$\angle c_n = \tan^{-1} \frac{0}{(1/n\pi)(-1)^{(n-1)/2}}. \tag{1.7.12}$$

Again, we must proceed carefully in determining the appropriate angles from Eq. (1.7.12). For $n = 1, 5, 9, 13, \ldots$, the denominator of the arctangent argument is positive, and hence from Fig. 1.6, the required angle lies in quadrant I or IV of the complex plane. Since the numerator is zero, we have immediately that $c_n = 0$ for these values of n. For $n = 3, 7, 11, 15, \ldots$, the denominator is negative, which locates the angle in the left half of the complex plane. Since the numerator is zero, $\angle c_n = \pm\pi$ radians for these values of n. By inspection for $n = 0$, $\angle c_n = 0$, and therefore to summarize,

$$\angle c_n = \begin{cases} 0°, & \text{for } n \text{ even} \\ 0°, & \text{for } n = 0 \\ 0°, & \text{for } n = 1, 5, 9, 13, \ldots \\ \pm 180°, & \text{for } n = 3, 7, 11, 15, \ldots \end{cases} \tag{1.7.13}$$

Notice that we associate an angle of zero degrees with those coefficients with a magnitude of zero.

Sketches of the amplitude and phase spectra of $f(t)$ given by Eqs. (1.7.11) and (1.7.13) are shown in Figs. 1.9 and 1.10. The phase is plotted as alternating between $\pm\pi$ since this seems to be a matter of convention. We could just as well have sketched the phase as $+\pi$ radians or $-\pi$ radians without alternating and still conveyed the same information.

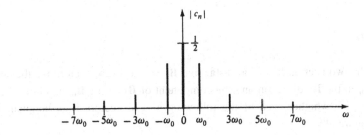

Fig. 1.9 Amplitude spectrum for Example 1.7.2

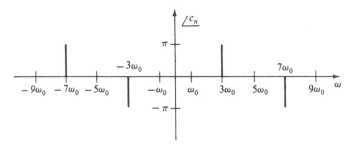

Fig. 1.10 Phase spectrum for Example 1.7.2

In both of the preceding examples, the real and imaginary parts of the complex coefficients could have been sketched instead of the amplitude and phase spectra without any loss of information. However, as noted previously, the amplitude and phase are more meaningful for our purposes, since they can be interpreted as the amplitude and phase of a sinusoid.

Summary

In this chapter we have expended substantial time and effort in motivating, defining, and developing the idea of a Fourier series representation of a signal or waveform. In Sect. 1.2 we provided the initial motivation and introduced sets of functions that possess the property of orthogonality. The two most important forms of Fourier series for our purposes, the trigonometric and the complex exponential forms, were presented in Sects. 1.3 and 1.4, respectively, and their use illustrated. In Sects. 1.5 and 1.6 we concentrated on simplifying the Fourier series calculations and supplying some previously ignored mathematical details. The important concept of a spectrum of a periodic signal was introduced in Sect. 1.7 and examples were given to clarify calculations. As we shall see, Sect. 1.7 serves as a critical stepping-stone to the definition of the Fourier transform in Chap. 2.

Problems

1.1 For the two vectors $\mathbf{B_0} = 3\hat{\mathbf{a}}_x + 4\hat{\mathbf{a}}_y$ and $\mathbf{B_1} = -\hat{\mathbf{a}}_x + 2\hat{\mathbf{a}}_y$, determine the component of $\mathbf{B_0}$ in the $\mathbf{B_1}$ direction and the component of $\mathbf{B_1}$ in the $\mathbf{B_0}$ direction.

1.2 Determine whether each of the following sets of vectors is orthogonal. Are they orthonormal?

(a) $\mathbf{B_0} = -\hat{\mathbf{a}}_x + \hat{\mathbf{a}}_y$ and $\mathbf{B_1} = \hat{\mathbf{a}}_x - \hat{\mathbf{a}}_y$.

(b) $\mathbf{B_0} = -\hat{\mathbf{a}}_x + \hat{\mathbf{a}}_y$ and $\mathbf{B_1} = -\hat{\mathbf{a}}_x - \hat{\mathbf{a}}_y$.

(c) $\mathbf{B}_0 = (\hat{\mathbf{a}}_x + \hat{\mathbf{a}}_y)/\sqrt{2}$ and $\mathbf{B}_1 = (\hat{\mathbf{a}}_x - \hat{\mathbf{a}}_y)/\sqrt{2}$.

1.3 We desire to approximate the vector

$$\mathbf{A}_1 = 4\hat{\mathbf{a}}_x + \hat{\mathbf{a}}_y + 2\hat{\mathbf{a}}_z$$

by a linear combination of the vectors $\mathbf{y}_1$ and $\mathbf{y}_2$,

$$\mathbf{A}_1' = d_1\mathbf{y}_1 + d_2\mathbf{y}_2,$$

where $\mathbf{y}_1 = -\hat{\mathbf{a}}_x + \hat{\mathbf{a}}_y$ and $\mathbf{y}_2 = -\hat{\mathbf{a}}_x - \hat{\mathbf{a}}_y$. Find the coefficients d_1 and d_2 such that the approximation minimizes the least squares loss function given by $\varepsilon^2 = |\mathbf{A}_1 - \mathbf{A}_1'|^2$.

1.4 Complete Example 1.2.1 by considering cases (2) and (3).

1.5 A set of functions $\{f_n(x)\}$ is said to be orthogonal over the interval (a, b) with respect to the weighting function $\rho(x)$ if the functions $\rho^{1/2}(x)f_n(x)$ and $\rho^{1/2}f_m(x), m \neq n$, are orthogonal, and thus

$$\int_a^b \rho(x)f_n(x)f_m(x)dx = 0.$$

The set of polynomials $\{H_n(x)\}$ defined by the equations

$$H_n(x) = (-1)^n e^{x^2/2} \frac{d^n}{dx^n} e^{-x^2/2}, \quad \text{for } n = 0, 1, 2, \ldots$$

are called *Hermite polynomials* and they are orthogonal over the interval $-\infty < x < \infty$ with respect to the weighting function $e^{-x^2/2}$. Specifically, this says that the two functions $e^{-x^2/4}H_m(x)$ and $e^{-x^2/4}H_n(x), n \neq m$, are orthogonal, so that

$$\int_{-\infty}^{\infty} e^{-x^2/2}H_m(x)H_n(x)dx = 0.$$

Show that $H_0(x)$ and $H_1(x)$ satisfy the relation above.

1.6 A set of polynomials that are orthogonal over the interval $-1 \leq t \leq 1$ without the use of a weighting function, called *Legendre polynomials*, is defined by the relations

$$P_n(t) = \frac{1}{2^n n!} \frac{d^n}{dt^n}(t^2 - 1)^n, \quad \text{for } n = 0, 1, 2, \ldots$$

and thus

$$P_0(t) = 1, \quad P_1(t) = t, \quad P_2(t) = \left(\frac{1}{2}\right)(3t^2 - 1),$$

and so on. Legendre polynomials are very closely related to the set of polynomials $\{1, t, t^2, \ldots, t^n, \ldots\}$ that we found to be nonorthogonal in Sect. 1.2. In fact, by using a technique called the Gram-Schmidt orthogonalization process [Jackson, 1941; Kaplan, 1959], the normalized Legendre polynomials, which are thus orthonormal, can be generated.

Calculate for the Legendre polynomials with $n = 0, 1, 2$, and so on, as necessary the value of

$$\int_{-1}^{1} P_n^2(t)dt$$

and hence use induction to prove that

$$\int_{-1}^{1} P_n^2(t)dt = \frac{2}{2n+1}.$$

Notice that this shows that the Legendre polynomials are not orthonormal. Based on these results, can you construct a set of polynomials that are orthonormal?

1.7 Show that the trigonometric functions in Example 1.2.1 are not orthonormal over $t_0 \le t \le t_0 + 2\pi/\omega_0$. From these results deduce a set of orthonormal trigonometric functions. Repeat both of these steps for the exponential functions in Example 1.2.2.

1.8 Notice that any polynomial in t can be expressed as a linear combination of Legendre polynomials. This fact follows straightforwardly from their definition in Problem 1.6, since we then have

$$1 = P_0(t), \quad t = P_1(t), \quad t^2 = \frac{2}{3}P_2(t) + \frac{1}{3}P_0(t), \quad t^3 = \frac{2}{5}P_3(t) + \frac{3}{5}P_1(t),$$

and so on. The reader should verify these statements. Using this result, then, we observe that the Legendre polynomial $P_n(t)$ is orthogonal to any polynomial of degree $n - 1$ or less over $-1 \le t \le 1$.. That is, for $g(t)$ a polynomial in t of degree $n - 1$ or less,

$$\int_{-1}^{1} P_n(t)g(t)\, dt = 0.$$

Demonstrate the validity of this claim for $n = 4$ and $g(t) = 5t^3 - 3t^2 + 2t + 7$.

1.9 Determine the trigonometric Fourier series expansion for the periodic function shown in Fig. P1.9 by direct calculation.

1.10 Obtain the trigonometric Fourier series representation of the waveform in Fig. P1.10.

1.11 Derive the magnitude-angle form of a trigonometric Fourier series given by

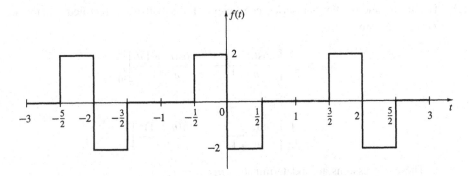

Fig. P1.9 Determine the trigonometric Fourier series expansion for the periodic function by direct calculation

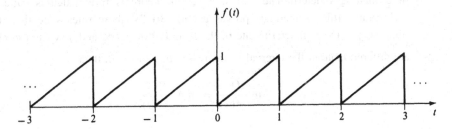

Fig. P1.10 Obtain the trigonometric Fourier series representation of the waveform

$$f(t) = a_0 + \sum_{n=1}^{\infty} d_n \cos\left[\frac{2\pi n t}{T} + \theta_n\right]$$

with

$$d_n = 2|c_n| = \sqrt{a_n^2 + b_n^2}$$

and

$$\theta_n = \tan^{-1}\frac{\text{Im}\{c_n\}}{\text{Re}\{c_n\}},$$

where the c_n are the complex Fourier series coefficients and the a_n and b_n are the trigonometric Fourier series coefficients. Start with Eq. (1.4.3) and use the fact that

$$c_n = |c_n|e^{j/\angle c_n} \quad \text{when } n > 0$$

and

$$c_n = |c_n|e^{-j/\angle c_n} \quad \text{when } n < 0$$

for $f(t)$ real.

1.12 The trigonometric Fourier series coefficients for a half-wave rectified version of $f(t) = \sin t$ are

$$a_n = \frac{1}{2\pi}\left[\frac{\cos(n-1)t}{n-1} - \frac{\cos(n+1)t}{n+1}\right]\Big|_0^\pi$$

and

$$b_n = \frac{1}{2\pi}\left[\frac{\sin(n-1)t}{n-1} - \frac{\sin(n+1)t}{n+1}\right]\Big|_0^\pi.$$

These expressions are indeterminate when $n = 1$.

(a) Use Eqs. (1.3.3) and (1.3.4) with $n = 1$ to show that $a_1 = 0$ and $b_1 = \frac{1}{2}$, respectively.

(b) In evaluating indeterminate forms, a set of theorems from calculus called L'Hospital's rules sometimes proves useful. Briefly, these rules state that if $\lim_{x \to x_0} f(x)/g(x)$ is indeterminate of the form $0/0$ or ∞/∞ and $f(x)$ and $g(x)$ are differentiable in the interval of interest with $g'(x) \neq 0$, then

$$\lim_{x \to x_0}\frac{f(x)}{g(x)} = \lim_{x \to x_0}\frac{f'(x)}{g'(x)}.$$

Use this rule to find a_1 and b_1 respectively. (See an undergraduate calculus book such as Thomas [1968] for more details on L'Hospital's rules).

1.13 Find the complex Fourier series representation of a nonperiodic function identical to $f(t)$ in Fig. 1.3.1 over the interval $-\tau/2 \leq t \leq \tau/2$.

1.14 Evaluate the complex Fourier series coefficients for the waveform in Fig. P1.9.

1.15 Determine the complex Fourier series representation for the waveform in Fig. P1.10.

1.16 Calculate the complex Fourier series coefficients for the periodic signal in Fig. P1.16.

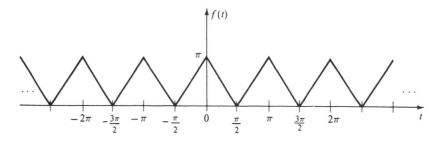

Fig. P1.16 Calculate the complex Fourier series coefficients for the periodic signal

1.17 Determine whether each of the waveforms in Fig. P1.17 is even, odd, odd harmonic, or none of these. Substantiate your conclusions.

1.18 For the waveforms specified below, determine if any of the symmetry properties discussed in Sect. 1.5 are satisfied. State any conclusions that can be reached concerning the trigonometric and complex Fourier series in each case.
   (a) Fig. P1.9
   (b) Fig. P1.10
   (c) Fig. P1.16

1.19 Prove that if $f(t) = -f(t - T/2)$, the complex Fourier series coefficients are zero for $n = 0, 2, 4, \ldots$

1.20 Prove that the integral of an odd function over symmetrical limits is zero; that is, show that if $f(t) = -f(-t)$, then

$$\int_{-a}^{a} f(t)dt = 0.$$

1.21 Obtain necessary conditions on the coefficients p_0, p_n, and q_n, $n = 1, 2, 3,\ldots$, in Eq. (1.6.3) to minimize Eq. (1.6.2). Do these coefficients have any special significance?

1.22 Obtain a trigonometric series approximation to $f(t)$ in Fig. P1.22 such that the total energy remaining in the error is 5% or less of the total energy in $f(t)$. Do not include any more terms than is necessary.

1.23 Plot the first three partial sums for the Fourier series of Fig. P1.22,

$$f(t) = \frac{1}{2} + \sum_{n=1}^{\infty} \left(\frac{-1}{n\pi}\right) \sin 2n\pi t,$$

on a large sheet of graph paper superimposed on the waveform in Fig. P1.22. What is the value of the integral squared error in this approximation?

1.24 What is the trigonometric series approximation to $f(t)$ in Fig. P1.9 such that the energy in the error is 15% or less of the total energy in $f(t)$? Include the fewest terms possible. Plot the resulting approximation over one period and compare to $f(t)$.

1.25 Derive Eq. (1.6.11).

1.26 Define $r_N(t) = h_N(t) - f_N(t)$,, and note that $f(t) - h_N(t) = f(t) - f_N(t) - r_N(t)$. Use this last expression in Eq. (1.6.2) to show that $f_N(t)$ is the one out of all possible trigonometric sums of order N that minimizes the ISE.

1.27 Derive the expression for the minimum value of the integral squared error in Eq. (1.6.14).

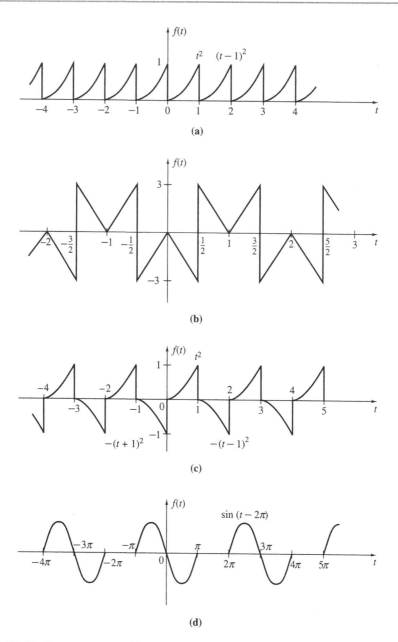

Fig. P1.17 Obtain necessary conditions on the coefficients p_0, p_n, and q_n, $n = 1, 2, 3,...$, in Eq. (1.6.3) to minimize Eq. (1.6.2). Do these coefficients have any special significance?

Fig. P1.22 Obtain a trigonometric series approximation to $f(t)$ in Fig. P1.22 such that the total energy remaining in the error is 5% or less of the total energy in $f(t)$. Do not include any more terms than is necessary.

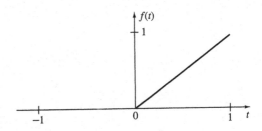

1.28 Approximate the waveform in Problem 1.13 by a series of Legendre polynomials of the form $f(t) = \sum_{n=0}^{2} \delta_n P_n(t)$. What percentage of the energy in $f(t)$ remains in the approximation error if we let $\tau = 1$?

1.29 Show that $\text{Re}\{c_n\}$ is even and $\text{Im}\{c_n\}$ is odd by finding the real and imaginary parts of $c_n = |c_n|e^{j/\angle c_n}$ and then using Eqs. (1.7.3) and (1.7.4).

1.30 Repeat Problem 1.29 by finding the real and imaginary parts of Eq. (1.4.4).

1.31 Find expressions for and sketch the amplitude and phase spectra of $f(t)$ in Fig. P1.9.

1.32 Repeat Problem 1.31 for $f(t)$ in Fig. P1.10.

1.33 Repeat Problem 1.31 for $f(t)$ in Fig. P1.16.

References

1. Jackson, D. 1941. *Fourier Series and Orthogonal Polynomials*. Washington, D.C.: The Mathematical Association of America.
2. Kaplan, W. 1959. *Advanced Calculus*. Reading, Mass.: Addison-Wesley.

Fourier Transforms

2

2.1   Introduction

As we saw in Chap. 1, trigonometric and complex exponential Fourier series are extremely useful techniques for obtaining representations of arbitrary periodic time waveforms. These methods also can be applied to nonperiodic signals if there is a specific time interval of interest outside which the accuracy of the representation is unimportant. No problems with this approach are evident until one compares the spectrum of a periodic signal and a nonperiodic signal expressed in this manner. We then quickly discover that since the two waveforms have the same Fourier series coefficients, they have the same spectrum! This is particularly disturbing since it is necessary that the spectrum of a function be unique, that is, not exactly the same as any other function, if the concept of a spectrum is to be of any utility to us. The reason for the occurrence of this problem is that the representation for the nonperiodic waveform was obtained somewhat artificially by assuming that the waveform was actually periodic for all time. It is our purpose in this chapter to develop a unique representation for nonperiodic signals that is valid for all time, $-\infty < t < \infty$, and therefore has a unique spectrum.

In Sect. 2.2 the Fourier transform pair, the Fourier transform and its inverse, is obtained from the complex exponential form of the Fourier series via a limiting argument. A sufficient condition for the existence of the Fourier transform is established in Sect. 2.3, and a special set of mathematical functions called generalized functions, which are very useful in communication systems analysis, are developed in Sect. 2.4. The Fourier transforms of certain signals involving generalized functions are defined in Sect. 2.5. Properties that can aid in the evaluation of Fourier transforms are stated and proven in Sect. 2.6. We complete the discussion of Fourier transforms in Sect. 2.7 by developing the concept of a spectrum for nonperiodic signals and by illustrating how the Fourier transform information can be presented in graphical form.

© The Author(s), under exclusive license to Springer Nature Switzerland AG 2023   31
J. D. Gibson, *Fourier Transforms, Filtering, Probability and Random Processes*,
Synthesis Lectures on Communications, https://doi.org/10.1007/978-3-031-19580-8_2

2.2   Fourier Transform Pair

There are numerous possible approaches for arriving at the required expressions for the Fourier transform pair. The approach used here is probably the simplest, most direct, and most transparent of any of these possibilities. We begin by considering slightly modified versions of the complex Fourier series and its coefficients given by

$$f_T(t) = \frac{1}{2\pi} \sum_{n=-\infty}^{\infty} c'_n e^{j\omega_n t} \Delta\omega_n \qquad (2.2.1)$$

and

$$c'_n = \int_{-T/2}^{T/2} f_T(t) e^{-j\omega_n t} \, dt, \qquad (2.2.2)$$

respectively. In Eqs. (2.2.1) and (2.2.2), $\omega_n = n\omega_0 = 2n\pi/T$, $\Delta\omega_n = \Delta(n\omega_0) = n\omega_0 - (n-1)\omega_0 = \omega_0$ is an incremental change in the frequency variable ω_n, and $c'_n = Tc_n$. With these definitions we see that Eqs. (2.2.1) and (2.2.2) are exactly equivalent to Eqs. (1.4.3) and (1.4.4) with $t_0 = -T/2$.

Holding the shape of $f_T(t)$ in the interval $-T/2 < t < T/2$ fixed, if we take the limit of $f_T(t)$ as $T \to \infty$, we obtain

$$f(t) = \lim_{T \to \infty} f_T(t) = \frac{1}{2\pi} \int_{-\infty}^{\infty} c'_n e^{j\omega t} \, d\omega \qquad (2.2.3)$$

and

$$c'_n = \int_{-\infty}^{\infty} f(t) e^{-j\omega t} \, dt, \qquad (2.2.4)$$

where we have let $\omega_n \to \omega$ and $\Delta\omega \to d\omega$. Noting that the complex Fourier series coefficients defined the discrete spectrum in this chapter, we define $F(\omega) = c'_n$, so Eqs. (2.2.3) and (2.2.4) become

$$f(t) = \frac{1}{2\pi} \int_{-\infty}^{\infty} F(\omega) e^{j\omega t} \, d\omega \qquad (2.2.5)$$

and

$$F(\omega) = \int_{-\infty}^{\infty} f(t)e^{-j\omega t}\, dt, \tag{2.2.6}$$

respectively.

Equation (2.2.6) is called the *Fourier transform* of the time function $f(t)$ and is sometimes denoted by $F(\omega) = \mathcal{F}\{f(t)\}$, while Eq. (2.2.5) is called the *inverse Fourier transform* of $F(\omega)$, sometimes written as $f(t) = \mathcal{F}^{-1}\{F(\omega)\}$. The two expressions together, Eqs. (2.2.5) and (2.2.6), are called the *Fourier transform pair*, and this relationship is sometimes indicated by $f(t) \leftrightarrow F(\omega)$.

A portion of the importance of the Fourier transform and its inverse can be attributed to the fact that the transform is *unique*. That is, every time function for which Eq. (2.2.6) is defined has a unique Fourier transform, and conversely, given its Fourier transform, we can exactly recover the original time function. This uniqueness property is critical, since without it, the transform would be useless. Of course, the Fourier transform possesses additional characteristics that give it advantages over other unique transforms. These characteristics are developed in the remainder of this chapter and Chap. 3.

The following two examples illustrate the calculation of the Fourier transform for two common time functions.

Example 2.2.1 Consider the signal $f(t)$ specified by

$$f(t) = \begin{cases} Ve^{-t/\tau}, & \text{for } t \geq 0 \\ 0, & \text{for } t < 0. \end{cases} \tag{2.2.7}$$

We desire to find the Fourier transform of $f(t)$ by using Eq. (2.2.6). By direct substitution, we have

$$F(\omega) = \int_{-\infty}^{\infty} f(t)e^{-j\omega t}\, dt = \int_{0}^{\infty} Ve^{-t/\tau}e^{-j\omega t}\, dt$$

$$= \frac{V\tau}{1 + j\tau\omega}. \tag{2.2.8}$$

A natural thing to do to complete the cycle would be to obtain $f(t)$ from $F(\omega)$ in Eq. (2.2.8) using Eq. (2.2.5). Plugging $F(\omega)$ into Eq. (2.2.5), we find that

$$f(t) = \frac{1}{2\pi} \int_{-\infty}^{\infty} \frac{V\tau}{1 + j\omega\tau}e^{j\omega t}\, d\omega, \tag{2.2.9}$$

which is a nontrivial integral. Of course, the integral in Eq. (2.2.9) can be evaluated using integral tables; however, this is not too instructive. Instead, we emphasize that since the Fourier transform is unique, once we find $F(\omega)$ given $f(t)$ in Eq. (2.2.7), we know that

Eq. (2.2.9) must equal Eq. (2.2.7) exactly.[1] Because of this fact and since extensive tables of the Fourier transform are available, we are not often required to evaluate difficult inverse transform integrals similar to Eq. (2.2.9).

Example 2.2.2 We desire to find the Fourier transform of the time function defined by

$$f(t) = \begin{cases} V, \text{ for } -\frac{\tau}{2} \le t \le \frac{\tau}{2} \\ 0, \text{ otherwise.} \end{cases} \tag{2.2.10}$$

From Eq. (2.2.6), we find immediately that

$$F(\omega) = \int_{-\tau/2}^{\tau/2} V e^{-j\omega t} \, dt = V\tau \frac{\sin(\omega\tau/2)}{\omega\tau/2}. \tag{2.2.11}$$

The reason for manipulating the result into this last form is explained in Sect. 2.7. Again, the inverse Fourier transform can be obtained by the use of Eqs. (2.2.5) and (2.2.11) and integral tables, or by inspection.

One final point should be made that $F(\omega)$ in each of Eqs. (2.2.8) and (2.2.11) is a continuous function of ω, as opposed to the discrete nature of the c_n in Chap. 1. As noted before, the expression of the Fourier transform in graphical form is considered in Sect. 2.7, after the Fourier transforms of many other common time functions are derived.

2.3   Existence of the Fourier Transform

Although we have obtained the general equations for the Fourier transform and its inverse in Sect. 2.2, we have not yet considered under what conditions the Fourier transform integral in Eq. (2.2.6) exists. This may seem like a strange statement to the reader. That is, what exactly do we mean when we say that an integral does or does not exist? We say that an integral *exists* or *converges* if

$$\int_a^b g(x) \, dx < \infty \tag{2.3.1}$$

and an integral does not exist or diverges if

---

[1] Depending on how *f(0)* is defined, this equality may not be exact at $t = 0$; however, such an occurrence has little physical significance (see Sect. 2.3).

$$\int_a^b g(x)\,dx = \infty. \tag{2.3.2}$$

Thus we see that in order for the Fourier transform to exist, we must have

$$\int_{-\infty}^{\infty} f(t)e^{-j\omega t}\,dt < \infty. \tag{2.3.3}$$

Since the integral in Eq. (2.3.3) is not always a simple one, we would like to have a condition that we can check without completely evaluating the Fourier transform of a function.

By using the Weierstrass M-test for integrals [1], it is possible to prove that the Fourier transform exists, or equivalently, Eq. (2.3.3) is true, if

$$\int_{-\infty}^{\infty} |f(t)|\,dt < \infty \tag{2.3.4}$$

[2] (see Problem 2.5). In words, a time function $f(t)$ has a Fourier transform if $f(t)$ is absolutely integrable, that is, if Eq. (2.3.4) holds. As we shall see, Eq. (2.3.4) is a fairly simple condition to test for and does not require that we calculate the Fourier transform completely. Actually, a couple of additional conditions are required to ensure convergence of the Fourier integral. These conditions are (1) that $f(t)$ must have a finite number of maxima and minima in any finite interval, and (2) that $f(t)$ must have a finite number of discontinuities in any finite interval. Under these last two conditions, the inverse Fourier transform converges to the average of $f(t)$ evaluated at the right- and left-hand limits at jump discontinuities. All functions that we consider will satisfy these last two requirements on maxima and minima and discontinuities.

Equation (2.3.4) and the last two conditions are sometimes called the *Dirichlet conditions*, and they are sufficient for the Fourier transform to exist, but they are not necessary. For instance, every time function that satisfies the Dirichlet conditions has a Fourier transform; however, there are some functions that have a Fourier transform but are not absolutely integrable. We shall encounter examples of such functions shortly.

Example 2.3.1 Using Eq. (2.3.4) to test $f(t)$ from Example 2.2.1, we find that for $V > 0$,

$$\int_{-\infty}^{\infty} |f(t)|\,dt = \int_0^{\infty} Ve^{-t/\tau}\,dt = -V\tau e^{-t/\tau}\Big|_0^{\infty} = V\tau < \infty$$

for V and τ finite. Since by inspection we see that $f(t)$ in Eq. (2.2.7) has a finite number of maxima and minima and a finite number of discontinuities in any finite interval, we conclude that the Fourier transform exists for $f(t)$

Notice that for the discontinuity at $t = 0$, the inverse Fourier transform converges to $[f(0+) + f(0-)]/2 = V/2$, which does not agree with Eq. (2.2.7) at $t = 0$. We can alleviate this discrepancy by letting $f(t = 0) = V/2$ in Eq. (2.2.7); however, since the value of $f(t)$ at precisely $t = 0$ is of little importance physically, we will not belabor the point.

Example 2.3.2 For $f(t)$ in Example 2.2.2 and $0 < V < \infty$,

$$\int_{-\infty}^{\infty} |f(t)| \, dt = \int_{-\tau/2}^{\tau/2} V \, dt = V\tau < \infty$$

for τ finite, which satisfies Eq. (2.3.4). Again, we see by inspection that the additional conditions are satisfied, and thus we conclude that $F(\omega) = \mathcal{F}\{f(t)\}$ exists for $f(t)$ as given.

The only two functions that we have tested thus far satisfy Eq. (2.3.4). Unfortunately, however, there are some very common and hence important signals that are not absolutely integrable. Two of these functions are considered in the following examples.

Example 2.3.3 The unit step function is defined by

$$u(t) = \begin{cases} 1, \text{ for } t \geq 0 \\ 0, \text{ otherwise,} \end{cases} \tag{2.3.5}$$

and is an extremely common and vital signal for testing system response. Directly, we discover that

$$\int_{-\infty}^{\infty} |f(t)| \, dt = \int_{0}^{\infty} (1) \, dt = \infty,$$

and therefore Eq. (2.3.4) is violated. The unit step function may still have a Fourier transform, but we cannot arrive at this conclusion from the Dirichlet conditions.

Example 2.3.4 The cosine function plays an important role in electrical engineering as a test signal and is used many times in communication systems as a carrier signal. However, it is easy to show that this function is not absolutely integrable; for *any* ω,

$$\int_{-\infty}^{\infty} |f(t)| \, dt = \int_{-\infty}^{\infty} |\cos \omega t| \, dt = (\infty) 2 \int_{-T/4}^{T/4} \cos \omega t \, dt,$$

where $T = 2\pi/\omega$ and since there are an infinite number of periods from $-\infty$ to $+\infty$. Continuing, we find that

$$\int_{-\infty}^{\infty} |f(t)| \, dt = (\infty) \frac{2}{\omega} \sin \omega t \Big|_{-T/4}^{T/4} = (\infty) \frac{4}{\omega} = \infty$$

and thus Eq. (2.3.4) is not satisfied. As in Example 2.3.3, it is therefore not possible to conclude whether or not a Fourier transform exists for this function.

Although Examples 2.3.3 and 2.3.4 indicate that we may have some difficulty in defining Fourier transforms for a step function and periodic waveforms, it is possible to obtain Fourier transforms for these important signals using the theory of generalized functions. We develop the necessary concepts in Sect. 2.4.

2.4 Generalized Functions

As mentioned previously, the unit step function and the sine and cosine functions are of fundamental importance to the study of electrical engineering. These functions are of no less import in the design and analysis of communication systems, and hence it is imperative that we be able to obtain Fourier transforms for these functions. Fourier transforms can be written for these and other functions that are not absolutely integrable by allowing the transforms to contain impulses or delta functions. The *impulse* or *delta function*, denoted by $\delta(t)$, is usually defined by the statements that

$$\delta(t) = 0, \quad \text{for } t \neq 0 \tag{2.4.1}$$

$$\delta(t) = \infty, \quad \text{for } t = 0 \tag{2.4.2}$$

and the function is infinite at the origin in the very special way that

$$\int_{-\infty}^{\infty} \delta(t) \, dt = 1. \tag{2.4.3}$$

Equations (2.4.1–2.4.3) are not consistent with the mathematics of ordinary functions, since any function that is zero everywhere except at one point cannot have a nonzero integral. A rigorous mathematical development of the delta function in the ordinary sense is thus not possible. However, a completely rigorous mathematical justification of the delta function is possible by using the theory of distributions developed by Schwartz [3] or the theory of generalized functions introduced by [4]. Because of the restricted scope of this book, a rigorous mathematical development will not be pursued here.

The delta function is usually considered to be the derivative of the familiar unit step function defined by Eq. (2.3.5). Of course, this relationship can be proven rigorously. However, rather than do so, we motivate the relationship between the impulse and unit step function by considering the unit step to be the limit of a finite ramp function with slope $1/\varepsilon$ as $\varepsilon \to 0$. The finite ramp is shown in Fig. 2.1. Thus we see that

$$u(t) = \lim_{\varepsilon \to 0} u_\varepsilon(t). \qquad (2.4.4)$$

If we take the derivative of $u_\varepsilon(t)$ with respect to t (at all points where the derivative exists), we obtain the function illustrated in Fig. 2.2. Notice that this function is a pulse of magnitude $1/\varepsilon$ and duration ε with an area of 1. As ε becomes smaller and smaller, the magnitude of the pulse becomes larger and the width decreases but the area remains constant at unity. Hence in the limit as $\varepsilon \to 0$ and $u_\varepsilon(t) \to u(t)$, the derivative of $u_\varepsilon(t)$ approaches an impulse, so we have obtained heuristically that

$$\delta(t) = \lim_{\varepsilon \to 0} \frac{d}{dt} u_\varepsilon(t). \qquad (2.4.5)$$

There are numerous interesting properties of the delta function that can be derived. Here we simply state without proof some of the more important ones for our purposes.

Fig. 2.1 Finite ramp function

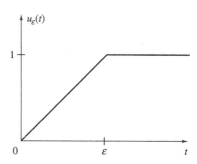

Fig. 2.2 Derivative of the finite ramp function

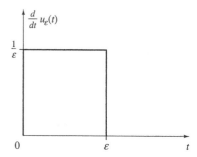

Sifting Property

$$\int_{-\infty}^{\infty} \delta(t - t_0)\phi(t)\,dt = \phi(t_0) \tag{2.4.6}$$

for $\phi(t)$ continuous at $t = t_0$.

The sifting property is probably the most used and most important property of the delta function. Notice that Eq. (2.4.3) is just a special case of Eq. (1.4.6), and therefore Eq. (2.4.6) is sometimes used to define a delta function.

The limits on the integral in Eq. (2.4.6) need not extend from $-\infty$ to $+\infty$ as long as the point $t = t_0$ lies within the limits of integration; that is, we can write

$$\int_{t_1}^{t_2} \delta(t - t_0)\phi(t)\,dt = \phi(t_0) \tag{2.4.7}$$

as long as $t_1 < t_0 < t_2$. If $t = t_0$ is outside the integration limits, the integral is zero.

Derivative Property

$$\int_{-\infty}^{\infty} \delta^{(n)}(t - t_0)\phi(t)\,dt = (-1)^n \phi^{(n)}(t)\Big|_{t=t_0} \tag{2.4.8}$$

for $\phi(t)$ continuous as $t = t_0$ and where the notation "superscript (n)" denotes the nth derivative of the function. The first derivative of the delta function $(n = 1)$ occurs somewhat frequently, and hence is given a special name, the *doublet*.

Scaling Property

$$\int_{-\infty}^{\infty} \delta(at)\phi(t)\,dt = \frac{1}{|a|}\phi(0). \tag{2.4.9}$$

Observe that with Eq. (2.4.6) this implies that

$$\int_{-\infty}^{\infty} \delta(at)\phi(t)\,dt = \frac{1}{|a|} \int_{-\infty}^{\infty} \delta(t)\phi(t)\,dt. \tag{2.4.10}$$

The following examples illustrate the application of the important properties developed in this section.

Example 2.4.1 We desire to evaluate the following integrals involving impulses:

(a) $\int\limits_{-\infty}^{\infty} \delta(t)[t + 1]\, dt$

(b) $\int\limits_{-\infty}^{\infty} \delta(t)e^{-3t}\, dt$

(c) $\int\limits_{-\infty}^{\infty} \delta(t - 2)[t^2 - 3t + 2]\, dt$

Solutions

(a) Using Eq. (2.4.6) with $\phi(t) = t + 1$, we find that

$$\int\limits_{-\infty}^{\infty} \delta(t)[t + 1]\, dt = [t + 1]|_{t=0} = 1.$$

(b) Again from Eq. (2.4.6) with $\phi(t) = e^{-3t}$, we obtain

$$\int\limits_{-\infty}^{\infty} \delta(t)e^{-3t}\, dt = e^{-3t}\big|_{t=0} = 1.$$

(c) Using the sifting property with $t_0 = 2$ and $\phi(t) = t^2 - 3t + 2$ yields

$$\int\limits_{-\infty}^{\infty} \delta(t - 2)[t^2 - 3t + 2]\, dt = 0.$$

Example 2.4.2 Evaluate the following integrals involving impulses:

(a) $\int\limits_{-2}^{1} \delta(t)\, dt$

(b) $\int\limits_{5}^{10} \delta(t - 7)e^{-t}\, dt$

(c) $\int\limits_{-3}^{-1} \delta(t)\left[t^2 + 2\right] dt$

(d) $\int\limits_{-3}^{0} \delta(t + 1)t^3 \, dt$

Solutions

(a) Directly from Eq. (2.4.6),

$$\int\limits_{-2}^{1} \delta(t) \, dt = 1.$$

(b) Using Eq. (2.4.7) with $t_1 = 5$, $t_0 = 7$, and $t_2 = 10$, we obtain

$$\int\limits_{5}^{10} \delta(t - 7)e^{-t} \, dt = e^{-7}.$$

(c) Here we have $t_0 = 0$, which lies outside the limits of integration; hence the integral is zero.

(d) We have $t_0 = -1$, so from Eq. (2.4.7),

$$\int\limits_{-3}^{0} \delta(t + 1)t^3 \, dt = t^3\big|_{-1} = -1.$$

Example 2.4.3 We desire to evaluate the following integrals using special properties of the impulse function.

(a) $\int\limits_{-\infty}^{\infty} \delta^{(1)}(t) \, dt$

(b) $\int\limits_{-\infty}^{\infty} \delta(2t)e^{-t/2} \, dt$

(c) $\int\limits_{-\infty}^{\infty} \delta^{(2)}(t - 1)\left[t^3 + t + 1\right] dt$

Solutions

(a) From the derivative property in Eq. (2.4.8) with $n = 1$, $t_0 = 0$, and $\phi(t) = 1$,

$$\int_{-\infty}^{\infty} \delta^{(1)}(t)\, dt = (-1)\frac{d}{dt}(1) = 0.$$

Notice that this is the integral of the doublet.

(b) Using the scaling property, Eq. (2.4.9), gives

$$\int_{-\infty}^{\infty} \delta(2t)e^{-t/2}\, dt = \frac{1}{|2|}e^{-t/2}\bigg|_{t=0} = \frac{1}{2}.$$

(c) A straightforward application of Eq. (2.4.8) produces

$$\int_{-\infty}^{\infty} \delta^{(2)}(t-1)\big[t^3 + t + 1\big]\, dt = (-1)^2\left\{\frac{d^2}{dt^2}\big[t^3 + t + 1\big]\right\}\bigg|_{t=1} = 6.$$

In this section we have introduced two special functions, the unit impulse and the unit step function, and have developed the properties of the impulse. Later we shall make liberal use of these functions and their properties.

2.5   Fourier Transforms and Impulse Functions

In this section we return to the problem of finding the Fourier transforms of functions that are not absolutely integrable. We consider the various singularity functions involved as limits of ordinary functions.[2] Although such an approach is nonrigorous, it allows the correct results to be obtained more easily and transparently.

Let us first calculate the Fourier transform of a delta function. By straightforward use of Eqs. (2.2.6) and (2.4.6), we find that

$$\mathcal{F}\{\delta(t)\} = \int_{-\infty}^{\infty} \delta(t)e^{-j\omega t}\, dt = e^{-j\omega t}\bigg|_{t=0} = 1. \tag{2.5.1}$$

Since Eq. (2.5.1) holds for all values of radian frequency ω, the delta function thus contains the same "amount" of all frequencies. Using the fact that the Fourier transform is unique, we have immediately from Eq. (2.5.1) that

---

[2] A singularity function is defined as a function that does not possess ordinary derivatives of all orders.

$$\mathcal{F}^{-1}\{1\} = \frac{1}{2\pi} \int_{-\infty}^{\infty} (1)e^{j\omega t}\, d\omega = \delta(t). \tag{2.5.2}$$

Hence we have the Fourier transform pair of the delta function,

$$\delta(t) \leftrightarrow 1. \tag{2.5.3}$$

Let us turn now to the calculation of the Fourier transform of a constant function of amplitude A that exists for all time, $-\infty < t < \infty$. Proceeding in a straightforward fashion, we have

$$\mathcal{F}\{A\} = \int_{-\infty}^{\infty} Ae^{-j\omega t}\, dt = \frac{-A}{j\omega} e^{-j\omega t} \Big|_{-\infty}^{\infty} = \frac{A}{j\omega}\left[e^{+j\infty} - e^{-j\infty}\right]. \tag{2.5.4}$$

Using Euler's identity on the exponentials in Eq. (2.5.4), we see that this integral does not converge; in fact, it oscillates continually for all values of ω. Note that we could have predicted this from Eq. (2.3.4), since a constant that exists for all time is not absolutely integrable.

To compute this Fourier transform, it is necessary to employ the unit impulse. From Eq. (2.5.2) we observe that

$$\int_{-\infty}^{\infty} e^{j\omega t}\, d\omega = 2\pi\delta(t). \tag{2.5.5}$$

Again writing the integral expression for the Fourier transform of A, we have

$$\mathcal{F}\{A\} = \int_{-\infty}^{\infty} Ae^{-j\omega t}\, dt = A \int_{-\infty}^{\infty} e^{-j\omega t}\, dt. \tag{2.5.6}$$

Making a change of variable in this last integral by letting $t = -x$ and using Eq. (2.5.5) yields

$$\mathcal{F}\{A\} = 2\pi A\delta(\omega). \tag{2.5.7}$$

Note that in Eq. (2.5.5), ω is simply a dummy variable of integration.

Another important function in the study of communication systems is the signum or sign function given by

$$\text{sgn }(t) = \begin{cases} 1, & \text{for } t > 0 \\ 0, & \text{for } t = 0 \\ -1, & \text{for } t < 0 \end{cases} \tag{2.5.8}$$

This function is also not absolutely integrable, and therefore an alternative method other than the direct integration of Eq. (2.2.6) is necessary to determine its Fourier transform. The approach taken here is to let

$$\text{sgn }(t) = \lim_{a \to 0} \left[e^{-a|t|} \text{sgn }(t) \right], \tag{2.5.9}$$

so that from Eq. (2.2.6) we obtain

$$\mathcal{F}\{\text{sgn }(t)\} = \mathcal{F}\left\{ \lim_{a \to 0} \left[e^{-a|t|} \text{sgn }(t) \right] \right\}$$

$$= \lim_{a \to 0} \int_{-\infty}^{\infty} e^{-a|t|} \text{sgn }(t) e^{-j\omega t} \, dt \tag{2.5.10}$$

upon interchanging the integration and limiting operations. Notice that we have not justi-fied mathematically the removal of the limit from under the integral sign. Although such a justification is required to ensure the validity of the derivation, we will not do so here, since the correct result is obtained in a direct and intuitive manner.

Returning to Eq. (2.5.10), we see that $e^{-a|t|}$ sgn (t) is absolutely integrable, and hence we proceed to evaluate the transform directly to find

$$\mathcal{F}\{\text{sgn }(t)\} = \lim_{a \to 0} \left\{ \frac{-1}{a - j\omega} + \frac{1}{a + j\omega} \right\} = \frac{2}{j\omega}. \tag{2.5.11}$$

The Fourier transform pair for the signum function is thus

$$\text{sgn }(t) \leftrightarrow \frac{2}{j\omega}. \tag{2.5.12}$$

If we observe that the unit step function can be written as

$$u(t) = \frac{1}{2} + \frac{1}{2}\text{sgn }(t), \tag{2.5.13}$$

it is now possible to determine the Fourier transform of $u(t)$ using Eqs. (2.5.7) and (2.5.12) and the linearity of the Fourier transform [see Eq. (2.6.6)] as

$$\mathcal{F}\{u(t)\} = \mathcal{F}\left\{\frac{1}{2}\right\} + \frac{1}{2}\mathcal{F}\{\text{sgn }(t)\} = \pi\delta(\omega) + \frac{1}{j\omega}. \tag{2.5.14}$$

Continuing the development of Fourier transforms for time functions that are not absolutely integrable, we consider the function $e^{j\omega_0 t}$ for $-\infty < t < \infty$. Although direct integration using Eq. (2.2.6) will not yield a convergent result, the desired Fourier transform can be calculated using an approach analogous to that employed in obtaining the transform of a constant. Substituting into Eq. (2.2.6) yields

$$\mathcal{F}\{e^{j\omega_0 t}\} = \int_{-\infty}^{\infty} e^{j\omega_0 t} e^{-j\omega t}\, dt = \int_{-\infty}^{\infty} e^{-j(\omega-\omega_0)t}\, dt. \tag{2.5.15}$$

By making the change of variable $y = -t$ and evaluating this integral by using Eq. (2.5.5), the Fourier transform of the complex exponential is found to be

$$\mathcal{F}\{e^{j\omega_0 t}\} = 2\pi\delta(\omega - \omega_0). \tag{2.5.16}$$

It remains for us to determine the Fourier transform of a sine or cosine function that exists for all time. Although back in Example 2.3.4 it seemed that this might be a sticky problem, we can now determine the transform almost trivially. Since $\cos \omega_0 t$ can be expressed as

$$\cos \omega_0 t = \frac{1}{2}\left[e^{j\omega_0 t} + e^{-j\omega_0 t}\right], \tag{2.5.17}$$

we find immediately from Eq. (2.5.16) that

$$\mathcal{F}\{\cos \omega_0 t\} = \pi[\delta(\omega - \omega_0) + \delta(\omega + \omega_0)]. \tag{2.5.18}$$

In a similar fashion we can show that the Fourier transform of $\sin \omega_0 t$ for $-\infty < t < \infty$ is

$$\mathcal{F}\{\sin \omega_0 t\} = j\pi[\delta(\omega + \omega_0) - \delta(\omega - \omega_0)]. \tag{2.5.19}$$

The reader may have noticed by this point that even though the Fourier transform originally was defined in Sect. 2.2 only for nonperiodic functions, we have been able to write transforms for certain periodic functions, namely the complex exponential, sine, and cosine functions, by using delta functions. The question, then, naturally arises as to whether it is possible to obtain the Fourier transform of any general periodic time function. The answer to this question is in the affirmative. The procedure is first to write the complex exponential Fourier series for the function of interest, and then take the Fourier transform of this series on a term-by-term basis. That is, for some general periodic function $f(t)$,

$$\mathcal{F}\{f(t)\} = \mathcal{F}\left\{\sum_{n=-\infty}^{\infty} c_n e^{jn\omega_0 t}\right\} = \sum_{n=-\infty}^{\infty} c_n \mathcal{F}\{e^{jn\omega_0 t}\}$$

$$= 2\pi \sum_{n=-\infty}^{\infty} c_n \delta(\omega - n\omega_0), \tag{2.5.20}$$

where the summation and integration can be interchanged since the series is uniformly convergent and the transform is a linear operation, and we have again used Eq. (2.5.5).

Example 2.5.1 As an illustration of the application of Eq. (2.5.20), let us find the Fourier transform of the periodic function shown in Fig. 1.1. From Example 2.3.1 we know that $\omega_0 = 2\pi/\tau$ and since we have previously shown that the c_n for this waveform are given by Eq. (2.4.6), we have that

$$\mathcal{F}\{f(t)\} = 2\pi \sum_{n=-\infty}^{\infty} c_n \delta\left(\omega - \frac{2n\pi}{\tau}\right) \tag{2.5.21}$$

with c_n as specified above.

An extremely important periodic signal for the analysis of communication systems is the train of unit impulses given by

$$\delta_T(t) = \sum_{n=-\infty}^{\infty} \delta(t - nT), \tag{2.5.22}$$

where T is the period. To find the Fourier transform of this function from Eq. (2.5.20), we first must find the complex Fourier series coefficients. Directly, using Eq. (1.4.4) we obtain

$$c_n = \frac{1}{T} \int_{-T/2}^{T/2} \delta_T(t) e^{-jn\omega_0 t} \, dt = \frac{1}{T} \int_{-T/2}^{T/2} \delta(t) e^{-jn\omega_0 t} \, dt = \frac{1}{T}. \tag{2.5.23}$$

Since $\omega_0 = 2\pi/T$, we have for the desired transform,

$$\mathcal{F}\{\delta_T(t)\} = \frac{2\pi}{T} \sum_{n=-\infty}^{\infty} \delta(\omega - n\omega_0). \tag{2.5.24}$$

Therefore, a train of unit impulses in the time domain with period T has as its Fourier transform a train of impulses with magnitude $\omega_0 = 2\pi/T$ spaced every ω_0 rad/sec. We will use this signal and its Fourier transform extensively in later analyses.

Transforms of several important time signals that are not absolutely integrable were obtained in this section by allowing the transforms to contain delta functions. Although the development of the transforms was nonrigorous, the results are nonetheless quite accurate, and we shall employ the Fourier transforms derived here time and again throughout our work.

2.6   Fourier Transform Properties

In calculating Fourier transforms and inverse transforms, several special situations and mathematical operations occur frequently enough that it is well worth our time to develop general approaches to handling them. In the following the most useful of the approaches are stated as properties of the Fourier transform pair, and a proof or method of proof is given for each property. When appropriate, an example of the application of the properties is presented immediately following each proof. The reader should note that when stating these properties, $g(\cdot)$ denotes a function of either time or frequency that has the shape g. Similarly, $G(\cdot)$ indicates a function that has the shape G, independent of whether it is a function of time or frequency. Hence $g(t)$ and $g(\omega)$ have the same shape, even though the first is a time signal and the latter is a function of frequency.

Symmetry Property
If

$$g(t) \leftrightarrow G(\omega),$$

then

$$G(t) \leftrightarrow 2\pi g(-\omega). \tag{2.6.1}$$

Proof: Use Eq. (2.2.5) and a change of variables.

Example 2.6.1 In Example 2.2.2 we found that

$$g(t) = \begin{cases} V, & \text{for } -\frac{\tau}{2} \leq t \leq \frac{\tau}{2} \\ 0, & \text{otherwise} \end{cases} \tag{2.6.2}$$

has the Fourier transform

$$G(\omega) = \mathcal{F}\{g(t)\} = V\tau \frac{\sin(\omega\tau/2)}{\omega\tau/2}. \tag{2.6.3}$$

Suppose now that we wish to find the Fourier transform of

$$G(t) = V \upsilon \frac{\sin (\upsilon t/2)}{\upsilon t/2}. \tag{2.6.4}$$

Direct calculation of $\mathcal{F}\{G(t)\}$ requires that we evaluate the integral

$$\mathcal{F}\{G(t)\} = \int_{-\infty}^{\infty} V \upsilon \frac{\sin (\upsilon t/2)}{\upsilon t/2} e^{-j\omega t} \, dt.$$

However, the symmetry property tells us from Eq. (2.6.1) that if $g(t)$ is known, then $\mathcal{F}\{G(t)\} = 2\pi g(-\omega)$. Using Eq. (2.6.2), we thus find that

$$\mathcal{F}\{G(t)\} = \begin{cases} 2\pi V, & \text{for } -\frac{\upsilon}{2} \le \omega \le \frac{\upsilon}{2} \\ 0, & \text{otherwise.} \end{cases} \tag{2.6.5}$$

Linearity Property.
   If

$$g_1(t) \leftrightarrow G_1(\omega)$$

and

$$g_2(t) \leftrightarrow G_2(\omega),$$

then for arbitrary constants a and b,

$$ag_1(t) + bg_2(t) \leftrightarrow aG_1(\omega) + bG_2(\omega). \tag{2.6.6}$$

Proof: Straightforward.
Scaling Property.
   If

$$g(t) \leftrightarrow G(\omega),$$

then for a real constant b,

$$g(bt) \leftrightarrow \frac{1}{|b|} G\left(\frac{\omega}{b}\right). \tag{2.6.7}$$

Proof: Use change of variables.

Example 2.6.2 Let $g_1(t) = g(t)$ given by Eq. (2.6.2), which has the Fourier transform $G_1(\omega) = G(\omega)$ as specified by Eq. (2.6.3). Let us find the Fourier transforms of the time functions

(i) $g_2(t) = g_1(2t) = \begin{cases} V, & \text{for } -\frac{\tau}{4} \le t \le \frac{\tau}{4} \\ 0, & \text{otherwise} \end{cases}$

and

(ii) $g_3(t) = g_1\left(\frac{t}{2}\right) = \begin{cases} V, & \text{for } -\tau \le t \le \tau \\ 0, & \text{otherwise} \end{cases}$

1. Notice that $g_2(t)$ is compressed in time when compared to $g_1(t)$. The Fourier transform of $g_2(t)$ can be found from $G'_i(\omega)$ and Eq. (2.6.7) by letting $b = 2$, so

$$\mathcal{F}\{g_2(t)\} = \frac{1}{2}G_1\left(\frac{\omega}{2}\right) = \frac{V\tau}{2}\frac{\sin(\omega\tau/4)}{\omega\tau/4}. \tag{2.6.8}$$

2. Here $g_3(t)$ is expanded in time compared to $g_1(t)$. Again using Eq. (2.6.7), this time with $b = \frac{1}{2}$, we have

$$\mathcal{F}\{g_3(t)\} = 2V\tau\frac{\sin(\omega\tau)}{\omega\tau}. \tag{2.6.9}$$

The scaling property is a very intuitive result, since a function that is compressed in time ($b > 1$) has a Fourier transform that extends to higher frequencies. Similarly, a function that is expanded in time varies more slowly with time and hence is compressed in terms of frequency content. A comparison of Eqs. (2.6.8) and (2.6.9) with $G_1(\omega)$ illustrates these last statements.

Time Shifting Property

If

$$g(t) \leftrightarrow G(\omega),$$

then

$$g(t - t_0) \leftrightarrow G(\omega)e^{-j\omega t_0}. \tag{2.6.10}$$

Proof: Straightforward change of variables.

Frequency Shifting Property
If

$$g(t) \leftrightarrow G(\omega),$$

then

$$g(t)e^{j\omega_0 t} \leftrightarrow G(\omega - \omega_0). \tag{2.6.11}$$

The notation $G(\omega - \omega_0)$ indicates the Fourier transform $G(\omega)$ centered about $+\omega_0$. Hence multiplication by $e^{j\omega_0 t}$ in the time domain produces a frequency shift of ω_0 radians per second in the frequency domain.

Proof: Direct substitution into Eq. (2.2.6).

The frequency shifting property is sometimes called the *modulation theorem*. The following example illustrates why this nomenclature is appropriate.

Example 2.6.3 To send information over long distances, it is necessary to translate the message signal to a higher frequency band. One way of accomplishing this translation is to form the product of the sum $s_0 + m(t)$, where s_0 is a constant and $m(t)$ is the message signal, and a carrier signal $\cos \omega_c t$. By using the frequency shifting property, it is possible to demonstrate that multiplication by $\cos \omega_c t$ does in fact achieve the desired frequency translation.

First working in the time domain, we have upon substituting the exponential form of $\cos \omega_c t$,

$$[s_0 + m(t)] \cos \omega_c t = s_0 \cos \omega_c t + m(t) \cos \omega_c t$$

$$= \frac{s_0}{2} \left[e^{j\omega_c t} + e^{-j\omega_c t} \right] + \frac{m(t)}{2} \left[e^{j\omega_c t} + e^{-j\omega_c t} \right]. \tag{2.6.12}$$

From Eq. (2.5.7) we know that

$$\mathcal{F}\left\{ \frac{s_0}{2} \right\} = \pi s_0 \delta(\omega) \tag{2.6.13}$$

and we denote by $M(\omega)$ the Fourier transform of the general message signal $m(t)$. Invoking the frequency shifting property, we find from Eq. (2.6.12) that

$$\mathcal{F}\{[s_0 + m(t)] \cos \omega_c t\} = \pi s_0[\delta(\omega - \omega_c) + \delta(\omega + \omega_c)]$$

$$+ \frac{1}{2}[M(\omega - \omega_c) + M(\omega + \omega_c)]. \qquad (2.6.14)$$

Since $M(\omega - \omega_c)$ and $M(\omega + \omega_c)$ denote the Fourier transform $M(\omega)$ centered about $+ \omega_c$ and $- \omega_c$, respectively, we see that the multiplication has had the desired result. The delta functions at $\pm\omega_c$ are also of great importance and their meaning is discussed in detail later.

Time Differentiation Property
If

$$g(t) \leftrightarrow G(\omega),$$

then

$$\frac{d}{dt} g(t) = j\omega G(\omega). \qquad (2.6.15)$$

Proof: Direct differentiation of the inverse Fourier transform.

Time Integration Property
If

$$g(t) \leftrightarrow G(\omega),$$

then

$$\int_{-\infty}^{t} g(\lambda) \, d\lambda \leftrightarrow \frac{1}{j\omega} G(\omega) + \pi G(0)\delta(\omega). \qquad (2.6.16)$$

Proof: The proof is not presented here, since it requires the use of the time convolution theorem presented in Sect. 3.3.

It is sometimes possible to use the special properties of the delta function and the time integration property to simplify greatly the evaluation of Fourier transforms. That is, since integrals involving impulses are quite easy to evaluate, if a function can be reduced by repeated differentiation to a sum of delta functions, the transform of this sum of impulses can be found and the time integration property employed to determine the transform of the original function. This procedure is examined in Problem 2.19.

In this section a host of important properties of the Fourier transform were stated and proven, and their application illustrated. These properties are used almost continually in the study of communication systems, and the student will soon have them committed to memory.

2.7   Graphical Presentation of Fourier Transforms

The Fourier transform of a time signal specifies the spectral content of the signal or the "amount" of each frequency that the signal contains. Thus far in this chapter we have concerned ourselves only with obtaining analytical expressions for Fourier transforms and have not discussed at all the graphical presentation of Fourier transform information. Since such graphical presentations can increase the understanding of Fourier transforms and many times constitute the most effective way of describing a signal's Fourier transform, we turn our attention to these graphical procedures in this section.

As mentioned in Example 2.2.1, the Fourier transform is, in general, a complex function, and hence two different graphs are necessary to present all of the information completely. Just as in the case of the complex Fourier series coefficients, we have a choice of which pair of graphs to plot. The real and imaginary parts are exactly that, the real and imaginary parts of the Fourier transform, and are denoted by $\text{Re}\{F(\omega)\}$ and $\text{Im}\{F(\omega)\}$, where $F(\omega)$ is given by Eq. (2.2.6). The *amplitude spectrum* is defined by the relation

$$| F(\omega) | = \left[(\text{Re}\{F(\omega)\})^2 + (\text{Im}\{F(\omega)\})^2\right]^{1/2} \qquad (2.7.1)$$

and the *phase spectrum* is given by

$$\angle F(\omega) = \tan^{-1} \frac{\text{Im}\{F(\omega)\}}{\text{Re}\{F(\omega)\}}. \qquad (2.7.2)$$

The amplitude and phase spectra are usually chosen over the real and imaginary parts for graphical presentation, since Eqs. (2.7.1) and (2.7.2) have a natural interpretation as the amplitude and phase, respectively, of an elementary sinusoid with radian frequency ω. Before proceeding, let us illustrate the calculation and graphical presentation of the amplitude and phase spectra.

Example 2.7.1 We wish to sketch the amplitude and phase spectra of $f(t)$ in Eq. (2.2.7). The Fourier transform for this signal was found in Example 2.2.1 to be

$$F(\omega) = \frac{V\tau}{1 + j\omega\tau} = \frac{V\tau(1 - j\omega\tau)}{1 + \omega^2\tau^2}. \qquad (2.7.3)$$

Directly from Eq. (2.7.4), we have

$$\text{Re}\{F(\omega)\} = \frac{V\tau}{1 + \omega^2\tau^2} \tag{2.7.4}$$

and

$$\text{Im}\{F(\omega)\} = \frac{-V\omega\tau^2}{1 + \omega^2\tau^2}. \tag{2.7.5}$$

Notice that in writing $\text{Im}\{F(\omega)\}$, the j is not included.

Substituting these last two results into Eq. (2.7.1), we obtain the amplitude spectrum of $F(\omega)$,

$$|F(\omega)| = \frac{V\tau}{\left[1 + \omega^2\tau^2\right]^{1/2}}. \tag{2.7.6}$$

From Eq. (2.7.2) we find for the phase spectrum,

$$\angle F(\omega) = \tan^{-1}\frac{-V\omega\tau^2/\left(1 + \omega^2\tau^2\right)}{V\tau/\left(1 + \omega^2\tau^2\right)} = \tan^{-1}(-\omega\tau) = -\tan^{-1}\omega\tau. \tag{2.7.7}$$

Sketches of Eqs. (2.7.6) and (2.7.7) are given in Figs. 2.3 and 2.4, respectively. These sketches contain all of the information in the analytical expression for $F(\omega)$, and by inspection of Figs. 2.3 and 2.4, we can determine the amplitude and phase associated with any frequency found in $f(t)$.

Just as was done in Sect. 1.5 for the Fourier series coefficients, it is also possible to prove various properties of $\text{Re}\{F(\omega)\}$ and $\text{Im}\{F(\omega)\}$ and the amplitude and phase spectra, depending on the form of $f(t)$. These properties are summarized in Table 2.1 and can simplify the sketching of the graphical presentation and can serve as a check on our analytical calculations. For instance, for purely real signals, which we are primarily interested in, we can show that the real part of $F(\omega)$ is an even function of ω and the imaginary part is odd. These facts can be quickly demonstrated by starting with Eq. (2.2.6), using Euler's identity, and working with the real and imaginary parts.

Fig. 2.3 Amplitude spectrum for Example 2.7.1

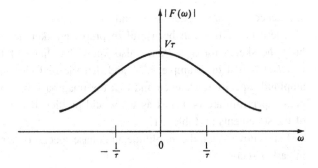

Fig. 2.4 Phase spectrum for
Example 2.7.1

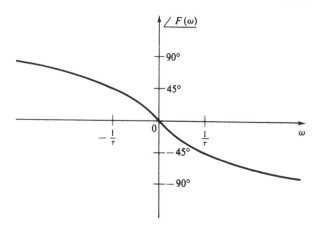

Table 2.1 Properties of the
Fourier Transform for Various
Forms of $f(t)$

If $f(t)$ is a:	Then F(ω) is a:
Real and even function of t	Real and even function of ω
Real and odd	Imaginary and odd
Imaginary and even	Imaginary and even
Complex and even	Complex and even
Complex and odd	Complex and odd

Considering now the amplitude spectrum of $F(\omega)$, we see by inspection of Eq. (2.7.1) that for $f(t)$ real, the amplitude spectrum is even since the square of an even or odd function is even. Note that we always assign the positive sign to the radical. Letting ω become $-\omega$ in Eq. (2.7.2) yields

$$\angle F(-\omega) = \tan^{-1} \frac{\text{Im}\{F(-\omega)\}}{\text{Re}\{F(-\omega)\}} = \tan^{-1} \frac{-\text{Im}\{F(\omega)\}}{\text{Re}\{F(\omega)\}}$$

$$= -\tan^{-1} \frac{\text{Im}\{F(\omega)\}}{\text{Re}\{F(\omega)\}} = -\angle F(\omega) \qquad (2.7.8)$$

and hence the phase spectrum is odd for $f(t)$ real.

These properties can be useful in preparing sketches of the spectra, since once we have the sketch for $\omega > 0$, we also know the shape of the spectra for $\omega < 0$. Notice that $f(t)$ is real in Example 2.7.1 and that the real part of its Fourier transform and its amplitude spectrum are even and that the imaginary part of its Fourier transform and its phase spectrum are odd, just as we would predict. It is not difficult to substantiate each of the statements in Table 2.1.

Let us now sketch the amplitude and phase spectra of some of the transforms calculated in earlier sections.

Fig. 2.5 Time waveform for
Example 2.7.2

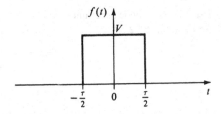

Example 2.7.2 Let us sketch the real and imaginary parts of the Fourier transform and the
amplitude and phase spectra for $f(t)$ given by Eq. (2.2.10) and shown in Fig. 2.5. The Fourier
transform was found in Example 2.2.2 to be

$$F(\omega) = V_\tau \, \frac{\sin \omega\tau/2}{\omega\tau/2}. \tag{2.7.9}$$

This Fourier transform is purely real and hence the real part is just $F(\omega)$ and the
imaginary part is zero. The real part of $F(\omega)$ is sketched in Fig. 2.6. A function of the
form of Eq. (2.7.9), which looks like the waveform in Fig. 2.6, is very common and is
sometimes called the sampling function. This is the reason for the extra manipulations in
Example 2.2.2 to obtain this form.

Only slightly more effort is required to obtain the amplitude and phase spectra. Since
$F(\omega)$ is purely real, the amplitude spectrum is

$$| F(\omega) | = V\tau \left| \frac{\sin \omega\tau/2}{\omega\tau/2} \right| \tag{2.7.10}$$

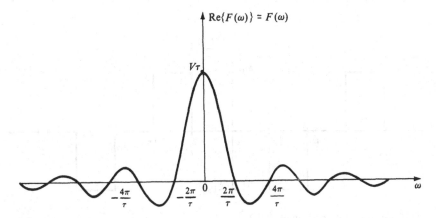

Fig. 2.6 $F(\omega)$ for Example 2.7.2

Fig. 2.7 Amplitude spectrum
for Example 2.7.2

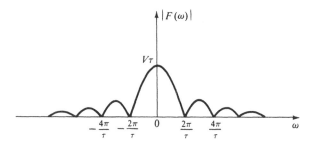

for $V > 0$, and it is sketched in Fig. 2.7. Although the phase spectrum can almost be
obtained by inspection, it is instructive to perform the required manipulations carefully.
Using Eqs. (2.7.2) and (2.7.11), we have

$$\angle F(\omega) = \tan^{-1} \frac{0}{\text{Re}\{F(\omega)\}}. \tag{2.7.11}$$

From Eq. (2.7.11) it is tempting to conclude that the phase is zero for all ω; however,
recall from similar situations in Chap. 1 that we must ascertain the sign of the denominator
in order to determine the angle correctly.

Since $\text{Re}\{F(\omega)\} = F(\omega)$, we see that the desired phase is

$$\angle F(\omega) = \begin{cases} \tan^{-1} \frac{0}{+1} = 0°, & \text{for } F(\omega) \geq 0 \\ \tan^{-1} \frac{0}{-1} = \pm 180°, & \text{for } F(\omega) < 0 \end{cases} \tag{2.7.12}$$

since only the sign of the denominator is important. The phase spectrum is sketched in
Fig. 2.8. The angle for $F(\omega) < 0$ is alternated between $+180°$ and $-180°$ by convention.

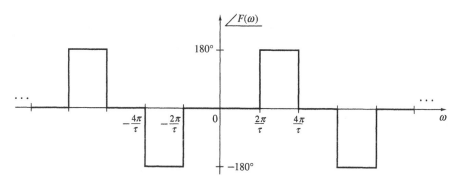

Fig. 2.8 Phase spectrum for Example 2.7.2

Fig. 2.9 Amplitude spectrum of cos $\omega_0 t$

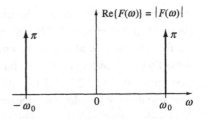

Fig. 2.10 Imaginary part of $\mathcal{F}\{\sin \omega_0 t\}$

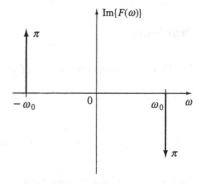

Example 2.7.3 Let us display graphically the Fourier transforms of cos $\omega_0 t$ and sin $\omega_0 t$. Working with cos $\omega_0 t$ first, we see from Eq. (2.5.18) that its Fourier transform is always real and positive, and thus Re$\{F(\omega)\} = |F(\omega)|$ and Im$\{F(\omega)\} = \angle F(\omega) = 0$. Hence only the one plot shown in Fig. 2.9 is necessary to convey all of the information.

The Fourier transform for sin $\omega_0 t$ is given by Eq. (2.5.19) and is purely imaginary. Therefore, only one plot is necessary again to display the Fourier transform information graphically, but this time the plot is of Im$\{F(\omega)\}$, which is sketched in Fig. 2.10.

The graphical display of Fourier transform information allows important comparisons to be made almost at a glance. The fundamental importance of the amplitude and phase spectra will become clearer as we progress through Chap. 3.

Summary

The Fourier transforms of a wide variety of time signals were calculated in this chapter and properties of the Fourier transform were presented to enable us to obtain those transforms not specifically given here. The definition of the Fourier transform pair and an intuitive development of the Fourier transform were given in Sect. 2.2. In Sect. 2.3 it was shown that a sufficient condition for the Fourier transform to exist is that the time function be absolutely integrable. After developing some special functions in Sect. 2.4, we were able to determine the Fourier transform of many important signals that are not abso-lutely integrable in Sect. 2.5 by allowing the transforms to contain impulses. In Sect. 2.6 we focused on several important properties that can greatly simplify the evaluation of some Fourier transforms. The techniques for presenting Fourier transforms graphically were given in Sect. 2.7 and the utility of these graphical methods was discussed and demonstrated. We employ the transforms and transform methods contained in this chapter continually throughout the remainder of the book.

Problems

2.1 Find the Fourier transform of the function

$$f(t) = \begin{cases} V, \text{ for } 0 \leq t \leq \tau \\ 0, \text{ otherwise} \end{cases}$$

for $0 < V < \infty$ by direct application of Eq. (2.2.6).

2.2 Determine the Fourier transform of the double-sided exponential function $f(t) = e^{-|t|/\tau}$ for $-\infty < t < \infty$ by direct application of Eq. (2.2.6).

2.3 Find the Fourier transform of the function $f(t)$ shown in Fig. P2.3.

2.4 Find the Fourier transform of the function $f(t)$

$$f(t) = \begin{cases} \frac{2V}{\tau}t, \text{ for } -\frac{\tau}{2} \leq t \leq \frac{\tau}{2} \\ 0, \quad \text{otherwise} \end{cases}$$

Fig. P2.3 Fourier transform of the function $f(t)$

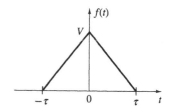

2.5 The Weierstrass M-test for integrals states that for $M(t)$ continuous in $a < t < \infty$ and $f(t, \omega)$ continuous in $a < t < \infty$ and $\omega_1 \leq \omega \leq \omega_2$ if

$$|f(t, \omega)| \leq M(t)$$

for $\omega_1 \leq \omega \leq \omega_2$ and

$$\int_a^\infty M(t)\, dt$$

converges, then

$$\int_a^\infty f(t, \omega)\, dt$$

is uniformly and absolutely convergent for $\omega_1 \leq \omega \leq \omega_2$ (Kaplan 1959). Starting with the Fourier transform integral in Eq. (2.2.6), use the M-test to show that Eq. (2.3.4) is sufficient for Eq. (2.3.3) to hold.

   Hint: Break the Fourier transform integral into a sum of two integrals and apply the M-test to each one.

2.6 Show that each of the following functions satisfies the Dirichlet conditions and hence has a Fourier transform: $f(t)$ in
   (a) Problem 2.1.
   (b) Problem 2.2.
   (c) Problem 2.3.
   (d) Problem 2.4.
2.7 Demonstrate that the function $f(t) = \sin \omega t$ for $\omega = 2\pi/T$ is not absolutely integrable.
2.8 In an heuristic fashion the waveform $\delta_\varepsilon(t)$ given in below Fig. P2.8a to approach a delta function in the limit as $\varepsilon \to 0$. Show that the derivative of $\delta_\varepsilon(t)$ approaches the function illustrated in below Fig. P2.8b, which is the doublet, as $\varepsilon \to 0$.

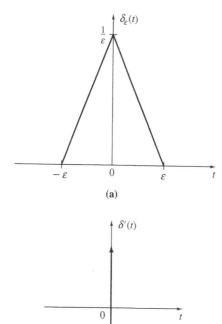

(a)

(b)

2.9 Evaluate the following integrals involving impulses.

(a) $\int\limits_{-\infty}^{\infty} \delta(t)t^3\, dt$

(b) $\int\limits_{0}^{\infty} \delta(t+1)e^{-t}\, dt$

(c) $\int\limits_{-2}^{5} \delta(t-4)[t+1]^2\, dt$

(d) $\int\limits_{-\infty}^{\infty} \delta(6t)e^{-3t}\, dt$

(e) $\int\limits_{-\infty}^{\infty} \delta^{(1)}(t-2)\left[t^3 - 3t^2 + 1\right] dt$

(f) $\int\limits_{-8}^{-1} \delta(t)\, dt$

(g) $\int\limits_{-\infty}^{6} \delta(t-1)e^{-t/2}\, dt$

2.10 Derive the Fourier transform of a constant A that exists for all time, given by Eq. (2.5.7), using an approach identical to that employed in obtaining $\mathcal{F}\{\text{sgn}(t)\}$ in Eqs. (2.5.10–2.5.12).

     Hint: Use L'Hospital's rule at $\omega = 0$.

2.11 Determine the Fourier transform of the periodic function in Fig. 1.2.

2.12 Find the Fourier transform of the periodic square wave in Fig. 1.5.

2.13 Use the results of Example 2.2.1 and the scaling property of the Fourier transform to find $\mathcal{F}\{e^{-t}u(t)\}$.

2.14 Determine the Fourier transform of $\delta(t - t_0)$ by using the time shifting property.

2.15 If the message signal $m(t)$ has the transform $M(\omega) = \mathcal{F}\{m(t)\}$, employ the frequency shifting property to find $\mathcal{F}\{m(t) \sin \omega_c t\}$.

2.16 Calculate the Fourier transform of $d(\sin \omega_0 t)/dt$ using Eq. (2.5.21) and the time differentiation property. Verify your result by the direct evaluation of $\mathcal{F}\{\omega_0 \cos \omega_0 t\}$.

2.17 Rework Problem 2.1 using the result of Example 2.2.2 and the time shifting property.

2.18 Determine the Fourier transform of the signal shown in the figure below using the linearity and time shifting properties of the Fourier transform.

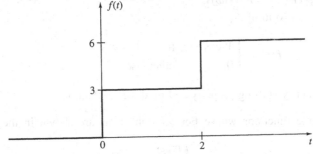

2.19 Compute the Fourier transform of the function

$$f(t) = \begin{cases} t + \frac{t^2}{2T}, & -T \le t < 0 \\ t - \frac{t^2}{2T}, & 0 \le t \le T \end{cases}$$

by using the linearity and time integration properties of the Fourier transform.

2.20 Show that the Fourier transform of the Gaussian pulse

$$f(t) = V e^{-t^2/2\tau^2}, \quad -\infty < t < \infty$$

is

$$F(\omega) = \tau V \sqrt{2\pi}\, e^{-\omega^2 \tau^2/2}, \quad -\infty < f < \infty,$$

which is also a Gaussian pulse.

2.21 Use the frequency shifting property and Eq. (2.2.8) to find the Fourier transform of

$$f(t) = e^{-t/\tau} \sin \omega_0 t \, u(t).$$

2.22 Determine the Fourier transform of the signal $f(t) = e^{-t/\tau} \cos \omega_0 t \, u(t)$. Let $\tau \to \infty$ and compare the resulting transform to that obtained by direct evaluation of $\mathcal{F}\{\cos \omega_0 t \, u(t)\}$.

2.23 Show that an arbitrary function can always be expressed as the sum of an even function $f_e(t)$ and an odd function $f_o(t)$. What are the even and odd components of $u(t)$?

2.24 Using the results of Problem 2.23 that any function can be written as $f(t) = f_e(t) + f_o(t)$, show that for $f(t)$ real (a) $\text{Re}\{F(\omega)\} = \mathcal{F}\{f_e(t)\}$, and (b) $j\text{Im}\{F(\omega)\} = \mathcal{F}\{f_o(t)\}$.

2.25 Show that $\mathcal{F}\{f^*(t)\} = F^*(-\omega)$.

   Hint: Let $f(t) = f_r(t) + jf_i(t)$, where $f_r(t)$ is the real part of $f(t)$ and $f_i(t)$ is the imaginary part (both purely real).

2.26 Use the result of Problem 2.25 to prove that for $f(t)$ a general complex function:
   (a) $\mathcal{F}\{f_r(t)\} = \frac{1}{2}\big[F(\omega) + F^*(-\omega)\big].$
   (b) $\mathcal{F}\{f_i(t)\} = \frac{1}{2j}\big[F(\omega) - F^*(-\omega)\big].$

2.27 Find the Fourier transform of

$$f(t) = \begin{cases} V \cos \omega_0 t, & \text{for } -\frac{\tau}{2} \le t \le \frac{\tau}{2} \\ 0, & \text{otherwise} \end{cases}$$

by using the frequency shifting property and by direct evaluation.

2.28 Determine the time functions whose Fourier transforms are shown in the figure below.

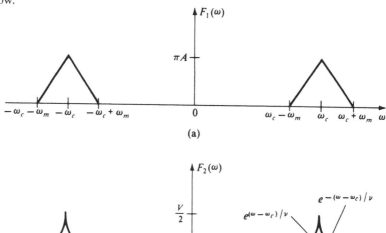

(a)

(b)

2.29 Sketch the amplitude and phase spectra for $f(t)$ in Problem 2.1.

2.30 Repeat Problem 2.29 for $f(t)$ in Problem 2.4.

References

1. Kaplan, W. 1959. *Advanced Calculus.* Reading, Mass.: Addison-Wesley.
2. Chen, W. H. 1963. *The Analysis of Linear Systems.* New York: McGraw-Hill
3. Schwartz, L. 1950. *Théorie des Distributions*, Vol. 1. Paris: Hermann.
4. Temple, G. 1953. "Theories and Applications of Generalized Functions." *J. London Math. Soc.,* Vol. 28, pp. 134–148.

Linear Systems, Convolution, and Filtering

3.1 Introduction

The representation of signals in both the time and frequency domains is extremely important, and this is exactly why two chapters have been devoted to this subject. Of equal importance, however, is the ability to analyze and specify circuits and systems that operate on, or process, signals to achieve a desired result. It is our purpose in this chapter to develop several techniques for system representation and analysis that are necessary for the study of the communication systems to follow.

We focus our investigation on the input/output behavior of systems and how this behavior can be expressed in the time and frequency domains. The discussion begins in Sect. 3.2, where we define what is meant by a linear system. A time-domain analysis method for calculating the output of a linear system for a given input, called convolution, is presented in Sect. 3.3, followed in Sect. 3.4 by a development of a graphical approach for performing the required convolution. The topic of filters and their use in signal processing is covered in Sects. 3.5 and 3.6, while the relationship between time response and system bandwidth is specified in Sect. 3.7. In Sect. 3.8 we discuss the important procedure of analog-to-digital conversion and the mathematical basis for such a procedure, the time-domain sampling theorem. Finally, the vital concepts of power and energy are defined in Sect. 3.9 for later use in system evaluation and comparison.

3.2 Linear Systems

All the discussions in this chapter are limited to linear time-invariant systems. Although this class of systems does not encompass all the systems that we will come in contact with while studying communication systems, there are several sound reasons for limiting

© The Author(s), under exclusive license to Springer Nature Switzerland AG 2023
J. D. Gibson, *Fourier Transforms, Filtering, Probability and Random Processes*,
Synthesis Lectures on Communications, https://doi.org/10.1007/978-3-031-19580-8_3

consideration to such systems. First, many physical systems are linear and time invariant or can be approximated accurately by linear time-invariant systems over some region of interest. Second, linear time-invariant systems are more easily analyzed than are other types of systems, and the analyses remain valid for very general conditions on the systems. Third, a thorough understanding of linear time-invariant systems is necessary before proceeding to study more advanced time-varying or nonlinear circuits and systems.

The adjectives *linear* and *time invariant* specify two separate and distinct properties of a system. We now define clearly what is meant by the terms *linear* and *time invariant*. There are numerous ways to determine whether a system is linear or nonlinear. One approach is to investigate the differential equation that represents the system. If all derivatives of the input and output are raised only to the first power and there are no products of the input and output or their derivatives, the system is said to be linear. An alternative definition requires only that we be able to measure or compute the output response for different inputs rather than the describing differential equation.

Definition Given that $y_1(t)$ and $y_2(t)$ are the system output responses to inputs $r_1(t)$ and $r_2(t)$ respectively, a system is said to be *linear* if the input signal $ar_1(t) + br_2(t)$ produces the system output response $ay_1(t) + by_2(t)$, where a and b are constants.

The reader may recognize this definition as the statement of a possibly more familiar concept called *superposition.*

The decision as to whether a system is time invariant or time varying can also be made by inspecting the system differential equation. If any of the coefficients in the differential equation are a function of time, the system is time varying. If the coefficients are all constants, the system is time invariant. Time invariance can also be established by observing the system output for a given input applied at different time instances, as stated by the following definition.

Definition Given that the output response of a system is $y_1(t)$ for an input $r(t)$ applied at $t = t_0$, a system is said to be *time invariant* if the output is $y_2(t) = y_1(t-t_1)$ for an input signal $r(t-t_1)$ applied at time $t = t_0 + t_1$.

In words, this definition says that a system is time invariant if the shape of the output response is the same regardless of when, in time, the input is applied.

Unless it is possible to measure the system response for various inputs in a laboratory, it is necessary that we work with a mathematical model of the system. There are four mathematical operations that are the basic building blocks of our mathematical models. These operations are scalar multiplication, differentiation, integration, and time delay, and the outputs for an input $r(t)$ applied at $t = t_0$ for systems that perform these operations are indicated as follows:

Scalar Multiplication

$$y(t) = \alpha r(t), \tag{3.2.1}$$

where α is an arbitrary constant.

Differentiation

$$y(t) = \frac{d}{dt} r(t) \tag{3.2.2}$$

Integration

$$y(t) = \int_{t_0}^{t} r(\tau) d\tau \tag{3.2.3}$$

Time Delay

$$y(t) = r(t - t_1) \tag{3.2.4}$$

Each of these operations describes a linear time-invariant system. The proofs of this statement are left as exercises (see Problems P3.1 and P3.2).

Let us investigate for a moment the implications of linearity or superposition. If a system is linear, the output response due to a sum of several inputs is the sum of the responses due to each individual input. Linear time-invariant systems also possess the significant property that the system output can contain only those frequencies that are present in the input. That is, no new frequencies are generated. The combination of the superposition and frequency preservation properties will prove to be of fundamental importance in determining the response of a linear system.

3.3 Linear Systems Response: Convolution

We now direct our attention toward the calculation of the output response of a linear time-invariant system to a given input waveform or excitation. We would like to keep the development as general as possible, so that the final result will be valid over a wide range of excitation signals and systems. Toward this end, consider the general waveform shown

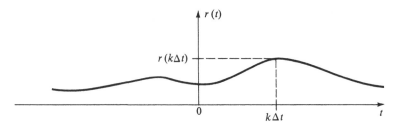

Fig. 3.1 General excitation signal

in Fig. 3.1, which is defined for $-\infty < t < \infty$. This signal is applied to the input of some general linear time-invariant system, and we wish to obtain an expression for the system output $y(t)$ in terms of $r(t)$ and some as yet unspecified characteristic of the system.

To do this, the signal $r(t)$ is written as an infinite sum of pulses of width Δt, as illustrated by Fig. 3.2. Limiting consideration to only the input pulse occurring at time $k\Delta t$ and assuming that the system response to such a pulse has the shape denoted by $h(\cdot)$, the output response to this one input pulse is

$$y(t) = r(k\Delta t)h(t - k\Delta t)\Delta t. \tag{3.3.1}$$

Equation (3.3.1) is the output due only to the pulse at $t = k\Delta t$ and consists of the magnitude of the input pulse times the response of the system to the pulse delayed by $k\Delta t$ multiplied by the duration of the input. Since the system is assumed to be linear, it obeys superposition and the total output can be obtained by summing the responses to each of the component pulses of $r(t)$. Hence we have

$$y(t) = \sum_{k=-\infty}^{\infty} r(k\Delta t)h(t - k\Delta t)\Delta t. \tag{3.3.2}$$

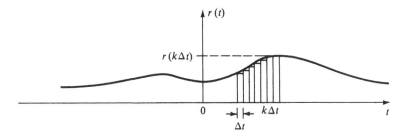

Fig. 3.2 Approximation of $r(t)$ by a sum of pulses

However, for the sequence of pulses of width Δt to represent $r(t)$ exactly, it is necessary that $\Delta t \to 0$. The final expression for the system output response is thus

$$y(t) = \lim_{\Delta t \to 0} \sum_{k=-\infty}^{\infty} r(k\Delta t)h(t - k\Delta t)\Delta t, \qquad (3.3.3)$$

which formally yields

$$y(t) = \int_{-\infty}^{\infty} r(\tau)h(t - \tau)d\tau. \qquad (3.3.4)$$

Equation (3.3.4) is an extremely important result and is usually called the *convolution integral*. Since Eq. (3.3.4) was derived by letting $\Delta t \to 0$, $h(\cdot)$ is thus the system response to an impulse input, and $h(t)$ is called the *impulse response* of the system being considered. Therefore, if the input signal and the system impulse response are known, the output can be calculated by the convolution operation indicated in Eq. (3.3.4). The symbol $*$ is commonly used as a shorthand notation for the convolution operation, and hence Eq. (3.3.4) can also be written as $y(t) = r(t) * h(t)$.

Before illustrating the use of the convolution integral, let us consider further the impulse response, $h(t)$. A more quantitative indication of why $h(t)$ is called the impulse response can be produced by returning to Eq. (3.3.4) and letting $r(\tau) = \delta(\tau)$, so that we are calculating the system output response to an impulse excitation. By applying the sifting property of the impulse, the output is found.

as

$$y(t) = \int_{-\infty}^{\infty} \delta(\tau)h(t - \tau)d\tau = h(t). \qquad (3.3.5)$$

Equation (3.3.5) makes the issue crystal clear; that is, $h(t)$ is called the impulse response because it is the system response to a unit impulse excitation.

One additional question concerning the convolution integral should be addressed. That is, Eq. (3.3.4) seems to be a very general expression for the output of a linear, time-invariant system, yet it depends on the system response to a particular input signal, the unit impulse. Does the impulse function have some property that uniquely qualifies it for the job, or could Eq. (3.3.4) just as well have been written in terms of the unit step response or response to some other excitation function? The answer is that the unit impulse does possess a special property that is evident from inspecting its Fourier transform in Eq. (2.5.1). The unit impulse contains all frequencies of equal magnitudes! Therefore, when we know the impulse response of a system, we have information concerning the system response to all frequencies. This fact emphasizes the importance of $h(t)$ in system response studies and further justifies its presence in Eq. (3.3.4).

Fig. 3.3 Circuit for Example
3.3.1

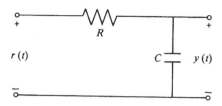

Example 3.3.1

The (voltage) impulse response of the simple RC network shown in Fig. 3.3 is given by

$$h(t) = \begin{cases} \frac{1}{RC}e^{-t/RC}, & \text{for } t > 0 \\ 0, & \text{for } t < 0. \end{cases} \tag{3.3.6}$$

We desire to determine the output response of this circuit to the input pulse

$$r(t) = \begin{cases} 1, & \text{for } 0 < t \le 2 \\ 0, & \text{otherwise.} \end{cases} \tag{3.3.7}$$

Using unit step functions to express $r(t)$ and $h(t)$, we have by direct substitution into Eq. (3.3.4) that

$$
\begin{aligned}
y(t) &= \int_{-\infty}^{\infty} [u(\tau) - u(\tau - 2)]\frac{1}{RC}e^{-(t-\tau)/RC}u(t - \tau)d\tau \\
&= \frac{1}{RC}\int_{-\infty}^{\infty} [u(\tau)u(t - \tau) - u(\tau - 2)u(t - \tau)]e^{-(t-\tau)/RC}d\tau \\
&= \frac{1}{RC}\int_{0}^{\infty} u(t - \tau)e^{-(t-\tau)/RC}d\tau - \frac{1}{RC}\int_{2}^{\infty} u(t - \tau)e^{-(t-\tau)/RC}d\tau. \tag{3.3.8}
\end{aligned}
$$

Notice that there are three different intervals for t which must be considered. For $t < 0$, $y(t) = 0$, since the factor $u(t-\tau)$ is zero for these values of t. For $0 \le t < 2$, Eq. (3.3.8) yields

$$y(t) = \frac{1}{RC}\int_{0}^{t} e^{-(t-\tau)/RC}d\tau = \left[1 - e^{-t/RC}\right]. \tag{3.3.9}$$

When $t \geq 2$, the output is given by

$$y(t) = \frac{1}{RC} \int_0^2 e^{-(t-\tau)/RC} d\tau = e^{-t/RC}\left[e^{2/RC} - 1\right]. \qquad (3.3.10)$$

Summarizing the results, we find that

$$y(t) = \begin{cases} 0, & \text{for } t < 0 \\ \left[1 - e^{-t/RC}\right], & \text{for } 0 \leq t < 2 \\ e^{-t/RC}\left[e^{2/RC} - 1\right], & \text{for } t \geq 2. \end{cases} \qquad (3.3.11)$$

Just as for the mathematical operation of multiplication, it is possible to show that the convolution operation is commutative, distributive, and associative. These properties are stated in the following paragraphs.

Commutative Law

$$r(t) * h(t) = h(t) * r(t) \qquad (3.3.12)$$

Proof: Use a change of variables.

Distributive Law

$$r(t) * [h_1(t) + h_2(t)] = r(t) * h_1(t) + r(t) * h_2(t) \qquad (3.3.13)$$

Proof: Straightforward.

Associative Law

$$r(t) * [h_1(t) * h_2(t)] = [r(t) * h_1(t)] * h_2(t) \qquad (3.3.14)$$

Proof: Deferred until later in this section.

Although we have a straightforward method of calculating the system response given the impulse response and the input signal, the evaluation of the convolution integral sometimes may be very challenging. In many cases, difficulties of this sort can be avoided by employing what is called the *time convolution theorem*.

Time Convolution Theorem

If

$$r(t) \leftrightarrow R(\omega)$$

and

$$h(t) \leftrightarrow H(\omega),$$

then

$$\mathcal{F}\{r(t) * h(t)\} = R(\omega)H(\omega). \tag{3.3.15}$$

Proof: From the definition of the Fourier transform,

$$\mathcal{F}\{r(t) * h(t)\} = \mathcal{F}\left\{ \int_{-\infty}^{\infty} r(\tau)h(t-\tau)d\tau \right\}$$

$$= \int_{-\infty}^{\infty}\left\{ \int_{-\infty}^{\infty} r(\tau)h(t-\tau)d\tau \right\}e^{-j\omega t}dt. \tag{3.3.16}$$

If we interchange the order of integration and make the change of variables $\lambda = t - \tau$, so that $d\lambda = dt$ in the inner integral,

$$\mathcal{F}\{r(t) * h(t)\} = \int_{-\infty}^{\infty} r(\tau)\left\{ \int_{-\infty}^{\infty} h(\lambda)e^{-j\omega(\lambda+\tau)}d\lambda \right\}d\tau$$

$$= \int_{-\infty}^{\infty} r(\tau)\left\{ \int_{-\infty}^{\infty} h(\lambda)e^{-j\omega\lambda}d\lambda \right\}e^{-j\omega\tau}d\tau$$

$$= H(\omega) \int_{-\infty}^{\infty} r(\tau)e^{-j\omega\tau}d\tau = R(\omega)H(\omega)$$

upon employing the definition of the Fourier transform twice.

Fig. 3.4 Two systems in cascade

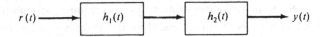

The time convolution theorem states that the Fourier transform of the convolution of two time functions is the product of their Fourier transforms. Therefore, if we know the impulse response of a system and we want to calculate the output response to a given input signal, one approach is to compute the Fourier transforms of the input and $h(t)$, form the product of these transforms, and then find the inverse transform. The power of the time convolution theorem is that frequently this last step is not necessary, since the frequency content of the response may be all the information required in many situations.

The time convolution theorem underscores the utility of the Fourier transform in systems analysis. For instance, suppose that it is desired to find the output response of two systems with impulse responses $h_1(t)$ and $h_2(t)$ that are connected in cascade as illustrated by Fig. 3.4. If the input signal is $r(t)$, we could, of course, use the associative law and perform two successive convolution operations. However, an alternative approach is simply to multiply the respective Fourier transforms of $r(t)$, $h_1(t)$, and $h_2(t)$ to obtain $Y(\omega) = \mathcal{F}\{y(t)\}$, and then, if necessary, take the inverse transform to get $y(t)$. We utilize such an approach repeatedly in the later chapters.

Just as it is possible to define the convolution operation in the time domain, it is also meaningful to consider the convolution of two signals in the frequency domain, as indicated by the following theorem.

Frequency Convolution Theorem

If

$$f(t) \leftrightarrow F(\omega)$$

and

$$g(t) \leftrightarrow G(\omega),$$

then

$$\mathcal{F}\{f(t)g(t)\} = \frac{1}{2\pi} F(\omega) * G(\omega). \tag{3.3.17}$$

Proof: Left as an exercise (see Problem 3.9).

Since most communication systems require the multiplication of two time signals, Eq. (3.3.17) is extremely useful for the analysis of these systems.

Let us now return to the proof of the associative law of convolution given by Eq. (3.3.14). Letting $\mathcal{F}\{r(t)\} = R(\omega)$, $\mathcal{F}\{h_1(t)\} = H_1(\omega)$, and $\mathcal{F}\{h_2(t)\} = H_2(\omega)$ and using the time convolution theorem to take the Fourier transform of the left-hand side of Eq. (3.3.14) yields

$$\mathcal{F}\{r(t) * [h_1(t) * h_2(t)]\} = R(\omega)[H_1(\omega)H_2(\omega)]$$
$$= [R(\omega)H_1(\omega)]H_2(\omega), \qquad (3.3.18)$$

where the last step is possible, since multiplication is associative. Taking the inverse transform of both sides of Eq. (3.3.18) establishes Eq. (3.3.14).

3.4   Graphical Convolution

As might be expected, graphical convolution consists of performing exactly those operations indicated by the convolution integral graphically. Consider the convolution of the two time functions $f_1(t)$ and $f_2(t)$ given by

$$f_1(t) * f_2(t) = \int_{-\infty}^{\infty} f_1(\tau) f_2(t - \tau) d\tau. \qquad (3.4.1)$$

The first operation involved in Eq. (3.4.1) is that of replacing t in $f_2(t)$ by $-\tau$, which constitutes a reflection about the $\tau = 0$ axis. The result of this first operation is then shifted by an amount of t seconds. Next, after letting $t = \tau$ in $f_1(t)$, the product of $f_1(\tau)$ and $f_2(t-\tau)$ is formed to obtain $f_1(\tau) \cdot f_2(t-\tau)$. Finally, this product is integrated over all values of τ, which is equivalent to computing the *area* under the *product*. These operations are repeated for all possible values of t to produce the total convolution waveform represented by Eq. (3.4.1). This last point must be emphasized. The analytical convolution in Eq. (3.4.1) yields a function that is defined for all values of t. Therefore, when performing the convolution graphically, it is necessary to calculate the area under the product for $-\infty < t < \infty$. To illustrate the concept of graphical convolution, consider the following brief example.

Example 3.4.1 We desire to convolve $f_1(t)$ graphically in Fig. 3.5 with $f_2(t)$ in Fig. 3.6. Replacing t in $f_2(t)$ with $-\tau$ and letting $t = \tau$ in $f_1(t)$, we are ready to begin the calculation. Shifting $f_2(-\tau)$ by an amount t and letting t increase from negative infinity, we find that the product $f_1(\tau) f_2(t - \tau) = 0$ for $-\infty < t < 0.5$.

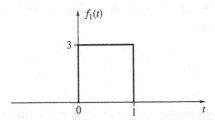

Fig. 3.5 $f_1(t)$ for Example 3.4.1

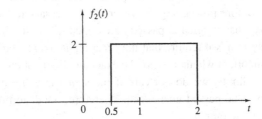

Fig. 3.6 $f_2(t)$ for Example 3.4.1

Further, over the interval $0.5 \le t \le 3$, $f_1(\tau)f_2(t-\tau) \ne 0$ while for $3 < t < \infty$, the product is again zero. The remaining details are left to the reader. The final result for all t is shown in Fig. 3.7.

Although there are situations when it is advantageous to perform the convolution graphically and there are occasions when the direct analytical approach is preferred, it often turns out that a combination of both methods is most efficient. The graphical approach can be used first to determine the values of t over which the convolution is nonzero and to get an idea of the shape of the resulting waveform. The convolution

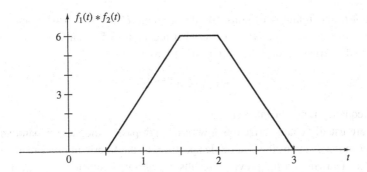

Fig. 3.7 $f_1(t) * f_2(t)$ for Example 3.4.1

can then be computed precisely using the limits deduced from the graphical analysis to evaluate the convolution integral.

3.5   Ideal Filters

The word *filter* is used by electrical engineers to denote a circuit or system that exhibits some sort of frequency selective behavior. Of course, every circuit or system fits this description to some extent; however, if for the specific application being considered all frequencies within the interval of interest are passed undistorted, the circuit or system is not acting as a filter. Before proceeding, however, it is necessary to specify exactly what we mean when we say that a signal is passed undistorted. For distortionless transmission through a circuit or system, we require that the exact input signal *shape* be reproduced at the output. It is not important whether or not the exact amplitude of the signal is preserved, and within reasonable limits, we do not care if the signal is delayed in time. Only the shape of the input must be passed unchanged. Therefore, for distortionless transmission of an input signal $r(t)$, the output is

$$y(t) = Ar(t - t_d). \tag{3.5.1}$$

Notice that Eq. (3.5.1) is a combination of the scalar multiplication and time-delay operations in Sect. 3.2, and can be shown to describe a linear time-invariant system.

Taking the Fourier transform of Eq. (3.5.1) with $\mathcal{F}\{y(t)\} = Y(\omega)$ and $\mathcal{F}\{r(t)\} = R(\omega)$ yields

$$Y(\omega) = AR(\omega)e^{-j\omega t_d}. \tag{3.5.2}$$

We know from the time convolution theorem that we also have

$$Y(\omega) = H(\omega)R(\omega), \tag{3.5.3}$$

where for $h(t)$ the impulse response of the system, $H(\omega) = \mathcal{F}\{h(t)\}$ and is usually called the *system transfer function*. By comparing Eqs. (3.5.2) and (3.5.3), we see that for distortionless transmission the system must have

$$H(\omega) = Ae^{-j\omega t_d} \tag{3.5.4}$$

over the frequency range of interest.

Filters are usually characterized as low pass, high pass, bandpass, or band stop, all of which refer to the shape of the amplitude spectrum of the filter's impulse response (transfer function). Drawing on the previous results concerning distortionless transmission, we find that an ideal low-pass filter (LPF) is defined by the frequency characteristic

$$H_{\text{LPF}}(\omega) = \begin{cases} Ae^{-j\omega t_d}, & \text{for } |\omega| \le \omega_s \\ 0, & \text{otherwise.} \end{cases} \tag{3.5.5}$$

The amplitude and phase of $H_{\text{LPF}}(\omega)$ are sketched in Figs. 3.8 and 3.9. Thus an ideal low-pass filter passes without distortion all input signal components with radian frequencies below ω_s, which is called the cutoff frequency. All signal components above the cutoff frequency are rejected.

The impulse response of the ideal LPF can be found by taking the inverse transform of Eq. (3.5.5), which yields

$$h_{\text{LPF}}(t) = \frac{A \sin \omega_s (t - t_d)}{\pi (t - t_d)} \tag{3.5.6}$$

by using the Fourier transform tables and the time shifting property. The impulse response has the form shown in Fig. 3.10 for $t_d \gg 1/2 f_s$. One example of an application of the ideal LPF is given in the following example.

Example 3.5.1 A signal $f(t)$ is applied to the input of an ideal LPF with $A = 2$, $t_d = 0$, and a cutoff frequency of $\omega_s = 40$ rad/sec. The magnitude of the transfer function of the filter is shown in Fig. 3.11. We would like to determine the output of this filter if the input signal

Fig. 3.8 Amplitude of the ideal low-pass filter transfer function

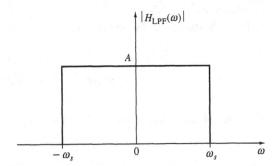

Fig. 3.9 Phase of the ideal low-pass filter transfer function

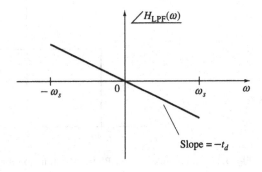

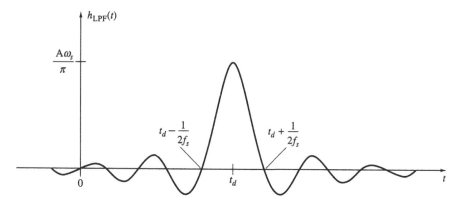

Fig. 3.10 Impulse response of the ideal LPF with $t_d \gg 1/2f_s$

is given by

$$r(t) = 3 + \sin 3t + \sin 12t - \cos 30t + 5\cos 47t$$
$$+ \sin 85t + 2\sin 102t + \cos 220t + \sin 377t. \tag{3.5.7}$$

Taking the Fourier transform of $r(t)$, we find that the amplitude spectrum of $r(t)$ is as illustrated in Fig. 3.12.

From the time convolution theorem, we know that if $y(t)$ is the output waveform, then

$$\mathcal{F}\{y(t)\} = Y(\omega) = H_{\mathrm{LPF}}(\omega)R(\omega)$$
$$= |H_{\mathrm{LPF}}(\omega)||R(\omega)|e^{j[\angle H_{\mathrm{LPF}}(\omega)+\angle R(\omega)]}. \tag{3.5.8}$$

If we are only interested in the magnitudes of the various components in the output, we have that

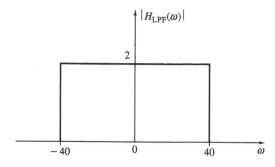

Fig. 3.11 Magnitude of the LPF transfer function for Example 3.5.1

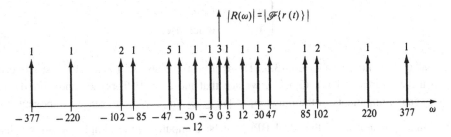

Fig. 3.12 Amplitude spectrum of the input for Example 3.5.1

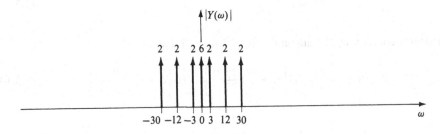

Fig. 3.13 Amplitude spectrum of the filter output for Example 3.5.1

$$|Y(\omega)| = |H_{\text{LPF}}(\omega)| \cdot |R(\omega)| \tag{3.5.9}$$

By performing the multiplication in Eq. (3.5.9) graphically, $|Y(\omega)|$ is found to be as shown in Fig. 3.13. Hence the output contains a dc term and components at ± 3, ± 12, and ± 30 rad/sec. All terms with frequencies above 40 rad/sec have been eliminated.

It is not possible to determine whether the impulses at ± 3, ± 12, and ± 30 rad/sec represent a sine or cosine function and whether the terms are positive or negative from Fig. 3.13 without also computing their phases. However, if we return to Eq. (3.5.8) and realize that since $t_d = 0$, $\angle H_{\text{LPF}}(\omega) = 0$, we see that the phase of each component is unchanged by the filter. The filter output can thus be found from Eq. (3.5.7) to be

$$y(t) = 6 + 2\sin 3t + 2\sin 12t - 2\cos 30t. \tag{3.5.10}$$

There are many situations in communication systems where it is necessary to pass signals that have frequency components between two nonzero frequencies, say ω_1 and ω_2, and reject all signals outside this range. A system that accomplishes this operation is called a bandpass filter. The transfer function of an ideal bandpass filter (BPF) is given by

$$H_{\text{BPF}}(\omega) = \begin{cases} Ae^{-j\omega t_d}, & \text{for } \omega_1 \leq |\omega| \leq \omega_2 \\ 0, & \text{otherwise.} \end{cases} \qquad (3.5.11)$$

The magnitude and phase of Eq. (3.5.11) are shown in Figs. 3.14 and 3.15, respectively. From these figures and Eq. (3.5.11), we see that the ideal BPF passes undistorted any input signal components in the range $\omega_1 \leq |\omega| \leq \omega_2$ and rejects all other components.

The impulse response of the ideal BPF can be computed quite simply by noting that $H_{\text{BPF}}(\omega)$ is just

$$H_1(\omega) = \begin{cases} Ae^{-j\omega t_d}, & \text{for } |\omega| \leq \frac{\omega_2 - \omega_1}{2} \\ 0, & \text{otherwise,} \end{cases} \qquad (3.5.12)$$

shifted in frequency by the amounts $\pm[(\omega_1 + \omega_2)/2]$. Hence

$$H_{\text{BPF}}(\omega) = H_1\left[\omega + \frac{\omega_1 + \omega_2}{2}\right] + H_1\left[\omega - \frac{\omega_1 + \omega_2}{2}\right]. \qquad (3.5.13)$$

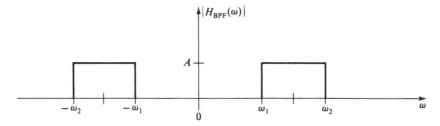

Fig. 3.14 Amplitude of the ideal bandpass filter transfer function

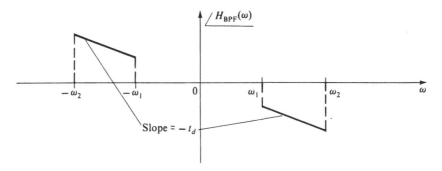

Fig. 3.15 Phase of the ideal bandpass filter transfer function

If we use the frequency shifting property and a table of Fourier transforms the impulse response of the ideal BPF can be found from Eq. (3.5.13) to be

$$h_{BPF}(t) = \frac{2A}{\pi} \cdot \frac{\sin\left[\left(\frac{\omega_2 - \omega_1}{2}\right)(t - t_d)\right]}{(t - t_d)} \cdot \cos\left[\frac{\omega_1 + \omega_2}{2}(t - t_d)\right]. \tag{3.5.14}$$

The magnitude and phase characteristics of an ideal high-pass filter (HPF) are shown in Figs. 3.16 and 3.17, respectively, and the entire transfer function is given by

$$H_{HPF}(\omega) = \begin{cases} Ae^{-j\omega t_d}, & \text{for } |\omega| > \omega_L \\ 0, & \text{otherwise.} \end{cases} \tag{3.5.15}$$

The ideal HPF rejects all input signal components at frequencies less than ω_L and passes all terms above ω_L with a multiplicative gain of A and a linear phase shift.

The last ideal filter we consider is the ideal band-stop filter (BSF), which has the amplitude and phase responses sketched in Figs. 3.18 and 3.19, respectively. As is evident from Fig. 3.18, this filter is designed to reject only those signals in a specified band

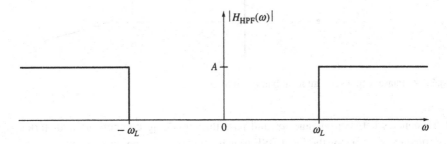

Fig. 3.16 Amplitude characteristic of an ideal high-pass filter

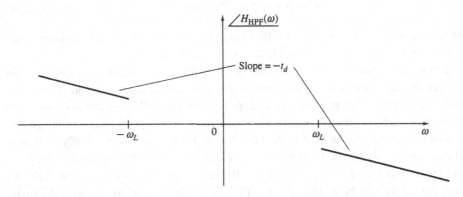

Fig. 3.17 Phase characteristic of an ideal high-pass filter

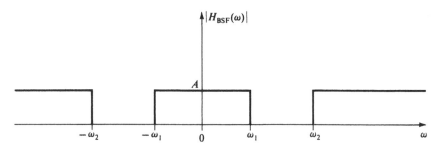

Fig. 3.18 Amplitude response of an ideal band stop filter

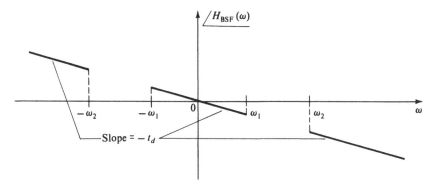

Fig. 3.19 Phase response of an ideal band stop filter

of frequencies between ω_1 and ω_2 and pass the remaining components undistorted. An analytical expression for the ideal BSF transfer function can be written as

$$H_{\text{BSF}}(\omega) = Ae^{-j\omega t_d} - H_{\text{BPF}}(\omega), \qquad (3.5.16)$$

where $H_{\text{BPF}}(\omega)$ is given by Eq. (3.5.13). Although we shall not have an occasion to use it, the impulse response of the ideal BSF can be found quite easily from Eq. (3.5.16).

It now seems that we have at our disposal virtually every kind of filter necessary for the design and analysis of communication systems, and hence we should be ready to leave the subject of filtering and proceed to the next topic. There is one very important detail that has been ignored. This detail is that the ideal filters discussed in the present section cannot actually be built in the lab. The ideal filters are not physically realizable. The reason for this is best illustrated by considering the ideal LPF that has the amplitude and phase responses in Figs. 3.8 and 3.9. If a unit impulse is applied to this LPF at $t = 0$, the output will be as given by Eq. (3.5.6), which is sketched in Fig. 3.10. Notice that the filter has a nonzero output *before* the impulse excitation signal is applied. Such a response is a physical impossibility. In the following section we investigate this situation in more detail and specify some physically realizable filters that adequately approximate the ideal filters in the present section.

3.6 Physically Realizable Filters

At the end of the preceding section, we reasoned that since the ideal LPF has a nonzero output prior to the application of an excitation signal, the filter cannot be constructed with physical components. This important condition, that is, the property that the system output cannot anticipate the input, is called the *causality condition.* A circuit or system is said to be *causal* or nonanticipatory if its impulse response $h(t)$ satisfies

$$h(t) = 0 \quad \text{for } t < 0. \tag{3.6.1}$$

By examining the impulse responses of the ideal LPF and BPF in Eqs. (3.5.6) and (3.5.14), respectively, both of these filters can be seen to violate Eq. (3.6.1). Additionally, since the ideal HPF and BSF transfer functions can be written in terms of the ideal LPF and BPF transfer functions, the impulse responses of these filters do not satisfy Eq. (3.6.1). All of these ideal filters are thus noncausal and hence not physically realizable.

Equation (3.6.1) expresses the causality requirement in the time domain. In terms of frequency-domain concepts, a system is said to be causal or *realizable* if

$$\int_{-\infty}^{\infty} \frac{|\log|H(\omega)||}{1 + \omega^2} d\omega < \infty, \tag{3.6.2}$$

where $H(\omega)$ is the system transfer function. Before applying Eq. (3.6.2), it is also necessary to establish that

$$\int_{-\infty}^{\infty} |H(\omega)|^2 d\omega < \infty. \tag{3.6.3}$$

The condition specified in Eq. (3.6.2) is called the *Paley-Wiener criterion.* Notice that the ideal LPF and BPF satisfy Eq. (3.6.3) and therefore the Paley-Wiener criterion can be used to determine the realizability of these filters. Both filters violate Eq. (3.6.2), however, since $|H_{\text{LPF}}(\omega)|$ and $|H_{\text{BPF}}(\omega)|$ are identically zero over a range of ω values. The ideal HPF and BSF transfer functions fail Eq. (3.6.3) and thus the Paley-Wiener criterion is not applicable to these filters.

Communication systems require the extensive use of all types of filters, and since it is not reasonable for an engineer to design a system using unrealizable elements, practical filters that achieve our goals must be found. Limiting consideration to the ideal LPF transfer function in Figs. 3.8 and 3.9, we see that there are three very stringent requirements that this filter satisfies:

1. Constant gain in the passband.
2. Linear phase response across the passband.
3. Perfect attenuation (total rejection) outside the passband.

We already know that it is not possible to realize a filter that exactly achieves all of these characteristics; and in fact, it is not possible to design a physically realizable filter that accurately approximates all three requirements. In practice, what has been done is to design three different types of filters, each of which provides a good approximation to one of the ideal LPF properties.

A type of practical filter called a *Butterworth filter* approximates the requirement of a constant gain throughout the passband. The amplitude characteristic of Butterworth low-pass filters can be expressed as

$$|H(\omega)| = \frac{1}{\left[1 + (\omega/\omega_c)^{2n}\right]^{1/2}},$$ (3.6.4)

where ω_c is the 3-dB cutoff frequency and $n = 1, 2, 3,...$ is the number of poles in the system transfer function, usually called the *order* of the filter. A sketch of Eq. (3.6.4) for $n = 1, 3$, and 5 and $\omega > 0$, along with the amplitude response of an ideal LPF, are given in Fig. 3.20. Notice that for $n = 1$ and ω_c appropriately defined in Eq. (3.6.4), we obtain the magnitude of the transfer function for the *RC* network in Fig. 3.3 (see Eq. 2.7.6).

Although Butterworth filters provide what is called a *maximally flat* amplitude response, their attenuation outside the desired passband may not be sufficient for many applications. A class of filters called *Chebyshev filters* provide greater attenuation for $\omega > \omega_c$ than Butterworth filters, and hence may prove useful in such situations. The amplitude response of an *n*th-order Chebyshev filter is given by

$$|H(\omega)| = \frac{1}{\left[1 + \varepsilon^2 C_n^2(\omega/\omega_c)\right]^{1/2}},$$ (3.6.5)

where $C_n(\cdot)$ denotes the *n*th-order Chebyshev polynomial from which the filter takes its name. Since Chebyshev polynomials are cosine polynomials, the amplitude response in Eq. (3.6.5) is not flat across the passband like the Butterworth filter, but contains ripples. The magnitude of the ripple depends on ε, since Eq. (3.6.5) oscillates between 1 and $\left[1 + \varepsilon^2\right]^{-1/2}$ for $0 \leq \omega/\omega_c \leq 1$.

Neither the Butterworth nor the Chebyshev filters exhibit a linear phase response across the passband, although the Butterworth approximation is not too bad. However, if true

Fig. 3.20 Butterworth and ideal LPF amplitude response

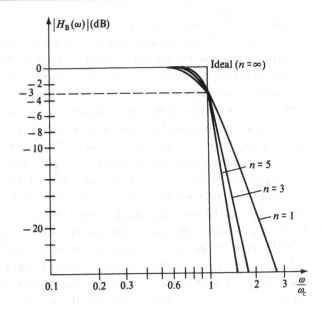

distortionless transmission of the signal phase is required, a class of filters called *Bessel filters* can be used. These filters, based on the Bessel polynomials, achieve a linear phase response at the expense of the other two ideal LPF requirements of constant gain across the passband and infinite attenuation outside the passband. The transfer function of a Bessel LPF is not written here, but it is pointed out that the Bessel LPF filters are obtained by truncating series expansions of the ideal LPF transfer function in Eq. (3.5.5).

Further discussion of filters and their design is not included here. For a more detailed development, the reader is referred to the many excellent books on the topic (see, e.g., [1]).

3.7   Time-Domain Response and System Bandwidth

The duality between the time and frequency domains is evident if one peruses the Fourier transform properties in Sect. 2.6. As a specific example, consider the time shifting and frequency shifting properties. This duality is important, since it sometimes provides intuition as to how a problem might be solved when otherwise there would be none. More important than the duality relationship, however, is the simple fact that for most signals and systems there exists both a time-domain and a frequency-domain representation. This fact, coupled with the time convolution theorem, allows us to perform analyses that would not be possible limited to either the time or frequency domain alone.

There are situations, unfortunately, where it may be difficult to transition from one domain to the other, and in these instances it is necessary to use information from one domain to make inferences concerning the other domain. One particularly common and crucial example of such a situation is the determination of system bandwidth from the system time response to some input. What is meant by the *bandwidth* of a system? For the present development we need only state that bandwidth is an indication of the range of real (positive) frequencies passed by the system, deferring a more quantitative definition until later in this section. Returning to the original discussion, there are many times in the laboratory when it is desired to obtain an approximate idea of a system's bandwidth. Even though it may be possible to observe on an oscilloscope the system impulse response (or a close approximation), it may be difficult to express $h(t)$ analytically so that $\mathcal{F}\{h(t)\} = H(\omega)$ can be calculated. Of course, another method of determining the system bandwidth is to plot a frequency response of the system. This approach is time consuming, and in some cases, may be impossible due to system restrictions. In these circumstances it may be possible to estimate the bandwidth from the system response to a step function input.

To be more explicit, let us calculate the response of the ideal LPF in Eqs. (3.5.5) and (3.5.6) to a unit step input. There are several ways to approach this problem, one of which is to employ the time convolution theorem and then find the inverse transform (Problem 3.23). Another method is to use the property that the output of a system due to the integral of some signal is the integral of the output due to the nonintegrated signal (Problem 3.24). This allows the step response to be found from the system impulse response (Problem 3.25). The approach taken here is that of direct computation using the convolution integral.

The step response of the ideal LPF in Eq. (3.5.6) is given by

$$y(t) = u(t) * h(t) = \int_{-\infty}^{\infty} \frac{A \sin \omega_s (\tau - t_d)}{\pi (\tau - t_d)} u(t - \tau) d\tau$$

$$= \int_{-\infty}^{t} \frac{A \sin \omega_s (\tau - t_d)}{\pi (\tau - t_d)} d\tau$$

$$= \int_{-\infty}^{t - t_d} \frac{A \sin \omega_s x}{\pi x} dx \tag{3.7.1}$$

upon making the change of variable $x = \tau - t_d$. The last integral in Eq. (3.7.1) cannot be evaluated in closed form, but the integral

$$\text{Si}(t) = \int_{0}^{t} \frac{\sin x}{x} dx \tag{3.7.2}$$

is tabulated extensively [2], and hence we manipulate Eq. (3.7.1) to get this last form. This produces

$$y(t) = \int_{-\infty}^{t-t_d} \frac{A}{\pi} \frac{\sin \omega_s x}{\omega_s x} \omega_s dx = \int_{-\infty}^{\omega_s(t-t_d)} \frac{A}{\pi} \frac{\sin v}{v} dv$$

$$= \frac{A}{\pi} \left\{ \text{Si}[\omega_s(t - t_d)] + \frac{\pi}{2} \right\}, \tag{3.7.3}$$

since $\text{Si}(-\infty) = -\pi/2$. The function $\text{Si}(t)$ in Eq. (3.7.2) has the shape shown in Fig. 3.21, and therefore the output $y(t)$ in Eq. (3.7.3) has the form sketched in Fig. 3.22.

Notice that in Fig. 3.22, the elapsed time from when the minimum of $y(t)$ occurs to t_d is $1/2f_s$ seconds, and the elapsed time from t_d until the maximum of $y(t)$ occurs is $1/2f_s$ seconds. These values can be deduced from a sketch of the $\sin \omega_s t/\omega_s t$ function since the minimum and maximum of the integral of this function occur at $t = -1/2f_s$ and $t = +1/2f_s$, respectively. If we define the *rise time* (t_r) as the time it takes for the system output to reach its maximum value from its minimum value, the rise time of the ideal LPF is $1/f_s$ seconds. Although this definition of rise time is somewhat arbitrary, any other definition simply changes the rise time by some multiplicative constant. The important point is that the system rise time to a unit step input is inversely proportional to the system bandwidth. This last statement remains true regardless of the definition of rise time. This relation between rise time and system bandwidth is very important, since

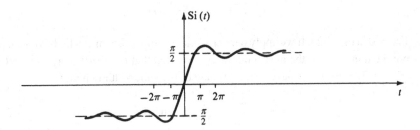

Fig. 3.21 Graph of $Si(t)$ in Eq. (3.7.2)

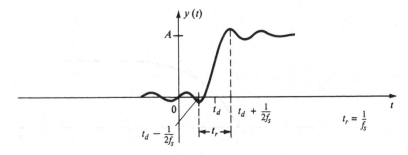

Fig. 3.22 Unit step response of an ideal low-pass filter

the output rise time to a unit step input is simple to compute in the lab even when a mathematical calculation or a frequency response plot is not feasible.

It is now necessary that we be more precise as to what is meant by the *bandwidth* of a system. If bandwidth (BW) is defined as the difference between the highest and lowest frequencies passed by a system, there is no difficulty in specifying the bandwidth of any of the ideal filters, since a frequency component is either passed undistorted or rejected. However, when attention is shifted to one of the practical filters, such as a Butterworth filter, the amplitude response is not perfectly flat within the desired band nor does it provide infinite attenuation outside this band of frequencies. Hence some quantitative definition is required. The almost universally accepted definition of bandwidth is the difference between the highest and lowest positive frequencies such that $|H(\omega)|$ is 0.707 of the magnitude at the filter's middle frequency (called *midband*). These upper and lower frequencies are generally called the *cutoff frequencies* of the filter. The following example illustrates the important concepts in this rather long-winded discussion.

Example 3.7.1 For the *RC* network in Fig. 3.3, let us first estimate the bandwidth from the output rise time to a unit step input, and then compare this result to the bandwidth computed from the system amplitude response. The unit step response of the network is easily shown to be

$$y(t) = \left[1 - e^{-t/RC}\right]u(t) \tag{3.7.4}$$

and is sketched in Fig. 3.23. If we apply our earlier definition of rise time, which is the elapsed time from the minimum to the maximum of $y(t)$, we find that $t_r \cong 10RC$ (say) $- 0 \cong 10RC$, since $e^{-10} \cong 0$. Therefore, we have as an estimate of the bandwidth in hertz:

$$\text{BW} \cong \frac{1}{t_r} = \frac{1}{10RC} \quad \text{hertz.} \tag{3.7.5}$$

The amplitude response of the *RC* network is available from Eq. (3.7.6) by letting $\tau = RC$ and $V \to 1/RC$ as

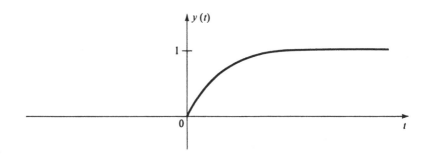

Fig. 3.23 Unit step response of the *RC* network in Fig. 3.3

$$|H(\omega)| = \frac{1}{\left[1 + \omega^2 R^2 C^2\right]^{1/2}}. \tag{3.7.6}$$

For a low-pass filter, midband is usually defined as $\omega = 0$, and thus we must determine the highest and lowest frequencies such that

$$|H(\omega)| = 0.707. \tag{3.7.7}$$

Since the lowest frequency passed by the filter is $\omega = 0$ and $|H(\omega = 0)| = 1 > 0.707$, $\omega = 0$ is the low-frequency value we are searching for. To find the higher frequency, we equate Eqs. (3.7.6) and (3.7.7) and solve for ω to produce $\omega = 1/RC$ rad/sec and $f = 1/2\pi RC$ hertz. The upper frequency is thus $1/2\pi RC$ hertz and the lower frequency is 0 Hz, and therefore,

$$BW = \frac{1}{2\pi RC} \quad \text{hertz.} \tag{3.7.8}$$

Comparing Eqs. (3.7.5) and (3.7.8), we see that the reciprocal of the rise time to a unit step response is indeed a reasonable estimate of the system bandwidth as we have defined it. Other definitions of rise time and bandwidth will change the accuracy of the estimate, but the basic reciprocal relationship between t_r and BW will remain valid.

Notice in Example 3.7.1 that the system bandwidth has a pronounced effect on the shape of the output waveform. That is, if the bandwidth of the system is narrow, then from Eq. (3.7.8), RC is large, and the rise time is very slow. As a result, the output given by Eq. (3.7.4) and sketched in Fig. 3.23 is very different from the input signal shape, $u(t)$. On the other hand, if the system has a wide bandwidth, RC is small and hence t_r is small. The output in this case greatly resembles the unit step input, and as $RC \to 0$ (BW $\to \infty$), $y(t) \to u(t)$. This same phenomenon is evident in Example 3.3.1.

3.8 The Sampling Theorem

Of all the theorems in communication theory, the one that is probably most often applied today is the time-domain sampling theorem. Electrical engineers, geologists, computer engineers, statisticians, and many other workers employ this theorem on a daily basis. The reason for the theorem's importance is that the digital computer is a powerful and commonly used tool for signal processing, and since we live in a world that is basically analog or continuous-time, some guidelines are necessary to allow analog signals to be digitized without loss of information. These guidelines are specified by the time-domain sampling theorem.

Time-Domain Sampling Theorem

If the Fourier transform of a time function is identically zero for $|\omega| > 2\pi B$ rad/sec, the time function is uniquely determined by its samples taken at uniform time intervals less than $1/2B$ seconds apart.

The sampling theorem thus states that any time function $f(t)$ with $\mathcal{F}\{f(t)\} = 0$ for $|\omega| > 2\pi B$ [such an $f(t)$ is usually said to be *bandlimited*] can be recovered exactly from its samples taken at a rate of $2B$ samples/sec or faster. A sampling rate of exactly $2B$ samples/sec is generally called the *Nyquist rate*. Although several proofs of the sampling theorem are possible, only one is presented here. This proof is probably the most illustrative of those available and is actually as much an example as it is a proof.

Proof: Given a bandlimited time signal, we desire to show that the signal can be uniquely recovered from its samples taken $1/2B$ seconds apart or less. Let $g(t)$ be a time signal bandlimited to B hertz, so that.

$$\mathcal{F}\{g(t)\} = G(\omega) = 0 \quad \text{for } \omega > 2\pi B, \tag{3.8.1}$$

where $G(\omega)$ is arbitrarily selected to be as shown in Fig. 3.24. Sampling $g(t)$ every T seconds by multiplying by the infinite train of impulses $\delta_T(t)$ in Eq. (2.5.22) yields

$$g_s(t) = g(t)\delta_T(t) = g(t) \sum_{n=-\infty}^{\infty} \delta(t - nT) = \sum_{n=-\infty}^{\infty} g(nT)\delta(t - nT) \tag{3.8.2}$$

with $T < 1/2B$.

Taking the Fourier transform of $g_s(t)$, we find that

$$\mathcal{F}\{g_s(t)\} = G_s(\omega) = \mathcal{F}\{g(t)\delta_T(t)\} = \frac{1}{2\pi}\mathcal{F}\{g(t)\} * \mathcal{F}\{\delta_T(t)\} \tag{3.8.3}$$

by the frequency convolution theorem in Eq. (3.3.17). Since $\mathcal{F}\{g(t)\} = G(\omega)$ and $\mathcal{F}\{\delta_T(t)\}$ is given by Eq. (2.5.24), we obtain for $\omega_0 = 2\pi/T$,

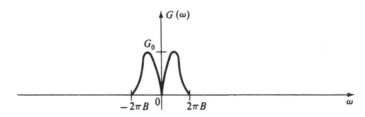

Fig. 3.24 $G(\omega)$ for proof of the sampling theorem

$$G_s(\omega) = \frac{1}{2\pi}G(\omega) * \omega_0 \sum_{n=-\infty}^{\infty} \delta(\omega - n\omega_0)$$

$$= \frac{\omega_0}{2\pi} \int_{-\infty}^{\infty} \left[\sum_{n=-\infty}^{\infty} G(\omega - v)\delta(v - n\omega_0) \right] dv$$

$$= \frac{1}{T} \sum_{n=-\infty}^{\infty} \left[\int_{-\infty}^{\infty} G(\omega - v)\delta(v - n\omega_0)dv \right]$$

$$= \frac{1}{T} \sum_{n=-\infty}^{\infty} G(\omega - n\omega_0) \tag{3.8.4}$$

by the sifting property of the impulse. The Fourier transform of $g_s(t)$ is therefore just $G(\omega)$ scaled by $1/T$ and centered about $\pm n\omega_0$. A sketch of $G_s(\omega)$ for $G(\omega)$ in Fig. 3.24 is sketched in Fig. 4.25 with $\omega_B = 2\pi B$.

We now have both time-and frequency-domain representations of the sampled time function $g(t)$, which are given by Eqs. (3.8.2) and (3.8.4), respectively. The question is whether or not $g(t)$ can be recovered undistorted from $g_s(t)$. The answer is "yes" if $g_s(t)$ is appropriately low-pass filtered. That is, if $g_s(t)$ is applied to the input of a ideal LPF with a cutoff frequency ω_s such that $\omega_B < \omega_s < \omega_0 - \omega_B$, only the center lobe of $G_s(\omega)$ in Fig. 3.25, which is simply $(1/T)G(\omega)$, will not be rejected. Since $\mathcal{F}^{-1}\{(1/T)G(\omega)\} = (1/T)g(t)$, we have indeed recovered $g(t)$ from $g_s(t)$ with only an amplitude change, and hence the sampling theorem is proved.

To add the finishing touches to the proof, let us consider the importance of the requirement that the signal be sampled at least every $1/2B$ seconds. First, did we make explicit use of this specification on the sampling rate in proving the theorem? Yes, we did, since we assumed that the period of $\delta_T(t)$ satisfies $T < 1/2B$. This in turn implies that

$$\omega_0 > 2\omega_B. \tag{3.8.5}$$

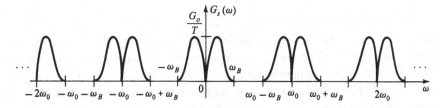

Fig. 3.25 $G_s(\omega)$ in Eq. (3.8.4) for $T < 1/2B$

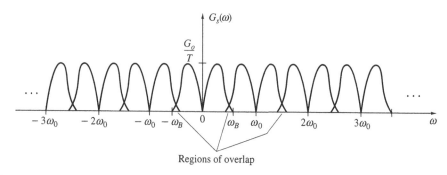

Fig. 3.26 $G_S(\omega)$ in Eq. (3.8.4) for $T \geq 1/2B$

Equation (3.8.5) thus says that the sampling frequency ω_0 must be greater than twice the highest frequency in $g(t)$, the signal being sampled. Why this restriction on sampling frequency is necessary becomes evident upon considering Fig. 3.25.

In this figure, the sketch indicates that $\omega_0 - \omega_B > \omega_B$ or $\omega_0 > 2\omega_B$, as we expect. What happens, however, if $\omega_0 \leq 2\omega_B$ (or equivalently, $T \geq 1/2B$) so that $\omega_0 - \omega_B \leq \omega_B$? When this situation occurs, $G_s(\omega)$ is as shown in Fig. 3.26. Notice the regions of overlap. Because of this overlap, it is no longer possible to recover $g(t)$ from $g_s(t)$ by low-pass filtering. In fact, it is not possible to obtain $g(t)$ from $g_s(t)$ by any means, since the frequency content in these regions of overlap adds, and therefore the signal is distorted. The distortion that occurs when a signal is sampled too slowly is called *aliasing*.

To complete the process begun in the proof, it is necessary to demonstrate how $g(t)$ is reconstructed from its samples by passing $g_s(t)$ through an ideal LPF. The exact calculation is left as an exercise. The final result is given by

$$g(t) = \sum_{n=-\infty}^{\infty} g(nT) \frac{\sin(\omega_s t - n\pi)}{\omega_s t - n\pi}, \tag{3.8.6}$$

which is displayed graphically in Fig. 3.27.

Example 3.8.1 With the increased availability of low-cost integrated circuits in recent years, digital voice terminals and digital transmission systems have become commonplace in the U.S. telephone network. The reason for this is simply that for some applications, the digital techniques provide improved performance at a lower cost than previous analog methods.

Since human speech is a continuous or analog signal, it is necessary that the speech be sampled before transmission through the digital systems. How often the speech signal must be sampled is dictated by the voice signal bandwidth and the sampling theorem. Since the frequency content of a speech signal can vary substantially depending on the speaker, the signal is low-pass filtered to some bandwidth less than 4 kHz prior to sampling. This bandlimited signal is then sampled 8000 times per second to satisfy the sampling theorem.

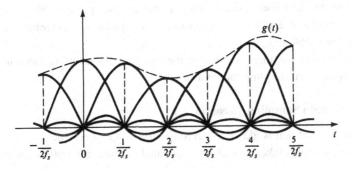

Fig. 3.27 $g(t)$ Reconstructed from its samples

The application of the sampling theorem described in Example 3.8.1 brings forth two important practical considerations. First, the sampling theorem requires that the time function be strictly bandlimited; that is, the Fourier transform of the function must be identically zero above some value of ω. Physical signals, however, are never absolutely bandlimited, since if a time signal is nonzero only over some finite interval, it can be shown that the signal contains components of all frequencies [see, e.g., Eqs. (2.2.10) and (2.2.11)]. Nonetheless, for all practical purposes it is reasonable to assume that most physical signals are bandlimited, since the majority of the signal "energy" (see Sect. 3.9) is contained in a finite frequency interval.

A second practical consideration is that even though a signal is low-pass filtered to a 3-dB bandwidth of, say, f_0 hertz, the signal is usually sampled at a rate greater than $2f_0$ samples/sec. For instance, as mentioned in Example 3.8.1, in the telephone network speech signals are low-pass filtered to less than 4 kHz (actually, bandpass filtered to 200–3400 Hz) but are sampled 8000 times a second, which is greater than the Nyquist rate of 6800 samples/sec. This is called *oversampling,* and it is, of course, permissible, since the sampling theorem is satisfied. However, why would one place an additional burden on the equipment by sampling faster than the Nyquist rate?

Part of the answer lies in the fact, illustrated by Fig. 3.25, that the higher the sampling rate, the more separation between adjacent lobes in the sampled signal's Fourier transform. The amount of the separation becomes critical when we realize two things: (1) The original analog signal is never perfectly band-limited, and (2) we must use a physically realizable filter, not an ideal one, to recover the original signal from its sampled version. Since a physical filter does not have perfect attenuation outside the passband, if the lobes are too close, part of the (unwanted) frequency content centered about $\pm\omega_0$ will also be recovered. This results in a distorted signal.

Another idealization employed in the proof of the sampling theorem is that of sampling with perfect impulses. Since in practice we cannot generate a true impulse or delta function, the sampling must be accomplished using finite width pulses. The question is: What effect, if any, will this have on recovering the message signal from the samples?

The answer to this question is that as long as the sampler pulse width is much smaller than the time between samples, the distortion is negligible. The derivation of this result is considered in Problem 3.37.

As one might surmise, there is a dual to the time-domain sampling theorem designated the frequency-domain sampling theorem.

Frequency-Domain Sampling Theorem

A time-limited signal that is identically zero for $|t| > T$ is uniquely determined by samples of its Fourier transform taken at uniform intervals less than π/T *rad/sec apart.*

Proof: Left as an exercise.

It should be noted that the sampling theorem discretizes the signal in time but not in amplitude. Thus, to complete the analog-to-digital conversion process, we must also discretize the amplitude of the samples.

3.9   Power and Energy

As the reader is well aware, power and energy are two very important concepts in the study of electrical circuits. These ideas also prove useful for signal and system analysis. In signal analysis, the *total energy* of a possibly complex time signal $f(t)$ in the time interval $t_1 < t < t_2$ is defined as

$$E = \int_{t_1}^{t_2} f(t)f^*(t)dt = \int_{t_1}^{t_2} |f(t)|^2 dt. \tag{3.9.1}$$

Although Eq. (3.9.1) is not dimensionally correct for energy in the electrical sense if $f(t)$ is a voltage, this definition is universal in signal and system analysis, since it allows a broader interpretation of the concept. If $t_1 = -\infty$ and $t_2 = +\infty$, Eq. (3.9.1) becomes the total energy in $f(t)$ for all time, given by

$$E = \int_{-\infty}^{\infty} |f(t)|^2 dt. \tag{3.9.2}$$

There are some signals for which Eq. (3.9.2) becomes infinite, and for these signals it is possible to define a related concept called the average power. The *average power* of a signal $f(t)$ in the time interval $t_1 < t < t_2$ is defined as

$$P = \frac{1}{t_2 - t_1} \int_{t_1}^{t_2} f(t)f^*(t)dt. \tag{3.9.3}$$

Equation (3.9.3) must be modified slightly to represent the average power over all time, which is given by

$$P = \lim_{\tau \to \infty} \frac{1}{2\tau} \int_{-\tau}^{\tau} f(t)f^*(t)dt. \tag{3.9.4}$$

The definitions of average power in Eqs. (3.9.3) and (3.9.4) can be motivated by recalling that "power measures the *rate* at which energy is transformed" (Smith 1971). Hence, if Eq. (3.9.1) represents the energy of $f(t)$ over the time interval $t_1 < t < t_2$, then Eq. (3.9.3) expresses the average rate of change in energy over the time interval. Of course, identical comments are appropriate for Eqs. (3.9.2) and (3.9.4). From Eqs. (3.9.2) and (3.9.4), it can be seen that a signal with finite energy has $P_\infty = 0$, while a signal with $E_\infty = \infty$ has finite average power over all time.

The following two examples illustrate the calculation of the energy and average power of a signal.

Example 3.9.1 It is desired to compute the energy over all time of the signal $f(t) = Ae^{-\alpha t}$ $u(t)$, where $\alpha > 0$. This signal is purely real, and hence, substituting into Eq. (3.9.2) produces

$$E_\infty = \int_{-\infty}^{\infty} A^2 e^{-2\alpha t} u(t)dt = \frac{A^2}{2\alpha} \tag{3.9.5}$$

for $-\infty < A < \infty$. By inspection of Eq. (3.9.4), we see that $P_\infty = 0$.

Example 3.9.2 We wish to compute the energy and average power over all time of the periodic signal $f(t) = A \sin \omega_c t$. From Eq. (3.9.2),

$$E_\infty = A^2 \int_{-\infty}^{\infty} \sin^2 \omega_c t\, dt = A^2 \int_{-\infty}^{\infty} \left[\frac{1}{2} - \frac{1}{2} \cos 2\omega_c t \right] dt = \infty. \tag{3.9.6}$$

The average power is, from Eq. (3.9.4),

$$P_\infty = \lim_{\tau \to \infty} \frac{A^2}{2\tau} \int_{-\tau}^{\tau} \sin^2 \omega_c t\, dt = \lim_{\tau \to \infty} \frac{A^2}{2\tau} \left[\frac{t}{2} - \frac{1}{4\omega_c} \sin 2\omega_c t \right] \Bigg|_{-\tau}^{\tau}$$

$$= \lim_{\tau \to \infty} \left[\frac{A^2}{2} - \frac{A^2}{4\omega_c \tau} \sin 2\omega_c \tau \right] = \frac{A^2}{2}. \tag{3.9.7}$$

As is suggested by this example, all periodic signals have $E_\infty = \infty$, but they may have P_∞ finite.

If energy and average power could be computed only by using Eqs. (3.9.1–3.9.4), their utility would be limited. This is because many situations occur where the frequency content of the signal is known, but not the time-domain waveform. To calculate the energy or average power would thus require that the inverse Fourier transform be taken first. Fortunately, frequency-domain expressions for the energy and average power over the infinite time interval can be obtained.

From the inverse Fourier transform relationship in Eq. (2.2.5), we have that

$$f^*(t) = \frac{1}{2\pi} \int_{-\infty}^{\infty} F^*(\omega)e^{-j\omega t} d\omega, \tag{3.9.8}$$

which when substituted into Eq. (3.9.2) yields

$$
\begin{aligned}
E_\infty &= \int_{-\infty}^{\infty} f(t) \left\{ \frac{1}{2\pi} \int_{-\infty}^{\infty} F^*(\omega)e^{-j\omega t} d\omega \right\} dt \\
&= \frac{1}{2\pi} \int_{-\infty}^{\infty} F^*(\omega) \left\{ \int_{-\infty}^{\infty} f(t)e^{-j\omega t} dt \right\} d\omega \\
&= \frac{1}{2\pi} \int_{-\infty}^{\infty} F^*(\omega) F(\omega) d\omega = \frac{1}{2\pi} \int_{-\infty}^{\infty} |F(\omega)|^2 d\omega.
\end{aligned}
\tag{3.9.9}
$$

Equation (3.9.9) is usually called *Parseval's theorem,* and it provides a method for calculating the energy in a signal directly from the signal's Fourier transform.

The quantity $|F(\omega)|^2$ under the integral in Eq. (3.9.9) is usually called the *energy density spectrum* of $f(t)$, since when it is multiplied by $1/2\pi$ and integrated over all ω, the total energy in $f(t)$ is obtained. The importance of the energy density spectrum and Eq. (3.9.9) is underscored by the following result. From Sect. 3.3 we know that the Fourier transform of the output of a linear system with transfer function $H(\omega)$ is given by $Y(\omega) = H(\omega)R(\omega)$, where $R(\omega)$ is the Fourier transform of the input. The energy density spectrum of the output is thus

$$
\begin{aligned}
|Y(\omega)|^2 &= Y(\omega)Y^*(\omega) = [H(\omega)R(\omega)]\left[H^*(\omega)R^*(\omega)\right] \\
&= |H(\omega)|^2 |R(\omega)|^2.
\end{aligned}
\tag{3.9.10}
$$

Therefore, given the system transfer function and the energy density spectrum of the input, the output energy density spectrum can be found from Eq. (3.9.10).

Example 3.9.3 Let us use Eq. (3.9.9) to calculate the energy in the time signal $f(t) = Ae^{-\alpha t}u(t)$ from Example 3.9.1. The Fourier transform of $f(t)$ is, from Example 2.2.1,

$$\mathcal{F}\{f(t)\} = \frac{A}{\alpha + j\omega} = F(\omega),$$

so that

$$|F(\omega)|^2 = \frac{A^2}{\alpha^2 + \omega^2}. \tag{3.9.11}$$

Substituting Eq. (3.9.11) into Eq. (3.9.9) yields

$$E_\infty = \frac{1}{2\pi} \int_{-\infty}^{\infty} \frac{A^2}{\alpha^2 + \omega^2} d\omega = \frac{A^2}{2\pi\alpha} \tan^{-1}\frac{\omega}{\alpha}\Big|_{-\infty}^{\infty} = \frac{A^2}{2\alpha}, \tag{3.9.12}$$

which agrees with the result of Example 3.9.1.

Since for some signals E_∞ is infinite, it is advantageous to obtain frequency-domain expressions for the average power of these signals, which are analogous to the energy density results in Eq. (3.9.9). Given a signal $f(t)$ for which E_∞ is infinite, we define a new time function

$$f_\tau(t) = \begin{cases} f(t), & |t| < \tau \\ 0, & \text{otherwise,} \end{cases} \tag{3.9.13}$$

which has the Fourier transform $F_\tau(\omega) = \mathcal{F}\{f_\tau(t)\}$. The average power of $f(t)$ is then given by

$$P_\infty = \lim_{\tau \to \infty} \frac{1}{2\tau} \int_{-\tau}^{\tau} |f(t)|^2 dt = \lim_{\tau \to \infty} \frac{1}{4\pi\tau} \int_{-\infty}^{\infty} |F_\tau(\omega)|^2 d\omega$$

$$= \frac{1}{2\pi} \int_{-\infty}^{\infty} \left\{ \lim_{\tau \to \infty} \frac{1}{2\tau} |F_\tau(\omega)|^2 \right\} d\omega \tag{3.9.14}$$

upon interchanging the integration and limiting operations. If the limit exists, the quantity in the braces in Eq. (3.9.14) is called the *power density spectrum,* and is usually denoted by $S_f(\omega)$.

The result for the power density spectrum analogous to Eq. (3.9.10) can be found by noting that the average output power over all time is

$$P_\infty = \lim_{\tau \to \infty} \frac{1}{2\tau} \int_{-\tau}^{\tau} |y(t)|^2 dt$$

$$= \frac{1}{2\pi} \int_{-\infty}^{\infty} |H(\omega)|^2 \left[\lim_{\tau \to \infty} \frac{|F_\tau(\omega)|^2}{2\tau} \right] d\omega. \tag{3.9.15}$$

The integrand in Eq. (3.9.15) is the output power density spectrum and the quantity in brackets is the input power density spectrum. Therefore, the output power density spectrum of a system with transfer function $H(\omega)$ can be found from the input power density spectrum using the equation

$$S_y(\omega) = |H(\omega)|^2 S_f(\omega). \tag{3.9.16}$$

It is pointed out that the expressions for the power density spectra in Eqs. (3.9.14) and (3.9.15) are valid only for deterministic signals, since interchanging the integration and limiting operations is not possible for random signals.

To illustrate the calculation of the average power of signals using Eq. (3.9.14), consider the following example.

Example 3.9.4 Let us employ Eq. (3.9.14) to calculate the average power over all time of $f(t) = A \sin \omega_c t$. The Fourier transform of the truncated signal is

$$F_\tau(\omega) = \int_{-\tau}^{\tau} A \sin \omega_c t e^{-j\omega t} dt = \frac{A}{j2} \int_{-\tau}^{\tau} \left[e^{-j(\omega - \omega_c)t} - e^{-j(\omega + \omega_c)t} \right] dt$$

$$= \frac{A\tau}{j} \left[\frac{\sin(\omega - \omega_c)\tau}{(\omega - \omega_c)\tau} - \frac{\sin(\omega + \omega_c)\tau}{(\omega + \omega_c)\tau} \right]. \tag{3.9.17}$$

Substituting $F_\tau(\omega)$ into Eq. (3.9.14) yields

$$P_\infty = \frac{1}{2\pi} \int_{-\infty}^{\infty} \left\{ \lim_{\tau \to \infty} \frac{1}{2\tau} \cdot A^2 \tau^2 \left[\frac{\sin(\omega - \omega_c)\tau}{(\omega - \omega_c)\tau} - \frac{\sin(\omega + \omega_c)\tau}{(\omega + \omega_c)\tau} \right]^2 \right\} d\omega$$

$$= \frac{A^2}{2} \int_{0}^{\infty} \left\{ \lim_{\tau \to \infty} \frac{\tau}{\pi} \left[\frac{\sin(\omega - \omega_c)\tau}{(\omega - \omega_c)\tau} \right]^2 + \lim_{\tau \to \infty} \frac{\tau}{\pi} \left[\frac{\sin(\omega + \omega_c)\tau}{(\omega + \omega_c)\tau} \right]^2 \right\} d\omega. \tag{3.9.18}$$

The cross-term falls out of Eq. (3.9.18) when the limit is taken, since the two sin x/x components become nonoverlapping.

One possible limiting expression for the unit impulse is

$$\delta(\omega) = \lim_{\tau \to \infty} \frac{\tau}{\pi} \left[\frac{\sin \omega \tau}{\omega \tau} \right]^2, \tag{3.9.19}$$

and hence Eq. (3.9.18) can be written as

$$P_\infty = \frac{A^2}{2} \int_0^\infty [\delta(\omega - \omega_c) + \delta(\omega + \omega_c)]d\omega = \frac{A^2}{2}, \qquad (3.9.20)$$

since for $\omega_c \neq 0$, only the first impulse falls within the limits of integration. Notice that Eq. (3.9.20) is the same value as that obtained in Example 3.9.2, which it must be.

Illustrations of the use of Eqs. (3.9.10) and (3.9.16) are not presented here, since the calculations required are much the same as demonstrated in Examples 3.9.3 and 3.9.4. The ability to compute the energy or average power will prove very important in later communication system comparative studies.

Summary

This chapter constitutes the final preparatory step before proceeding to the analysis of communication systems in the absence of noise. In Sects. 3.2 through 3.4, the concept of a linear time-invariant system was developed and time-and frequency-domain methods for determining the system output for a specified input were derived and illustrated. The input/output behavior of ideal and physically realizable filters was investigated in Sects. 3.5 and 3.6, followed in Sect. 3.7 by a discussion of system bandwidth and its relationship to the system time response. The important topics of analog-to-digital and digital-to-analog conversions were introduced in Sect. 3.8 with a development of the time-domain sampling theorem. Expressions for the calculation of the energy and average power of signals from both time-and frequency-domain representations were presented and illustrated in Sect. 3.9. Each of the ideas and concepts discussed in this chapter is employed repeatedly in the analyses that follow.

Problems

3.1 Prove that the differentiation operation in Eq. (3.2.2) describes a linear time-invariant system.

   Hint: For time-invariance, use the definition of the derivative.

3.2 Show that the time-delay operation in Eq. (3.2.4) represents a linear time-invariant system.

3.3 Find and sketch the output response of the RC network in Fig. 3.3 to the input signal $r(t) = \delta(t) + 2\delta(t-3) + \delta(t-4)$ if $R = 1$ MΩ and $C = 1$ μF.

3.4 For $RC = 1$, sketch the system response given by Eq. (3.3.11).

3.5 Use Eq. (3.3.4) to find the response of the RC network in Fig. 3.3 to the input signal
     $r(t) = \sin \omega_0 t$ for $-\infty < t < \infty$.

3.6 Evaluate the following convolution integrals.
     (a) $e^{-t}u(t) * e^{-2t}u(t)$.
     (b) $e^{-t}u(t) * t\, u(t)$.

3.7 Use Eq. (3.3.4) to compute $f_1(t) * f_2(t)$, where $f_1(t)$ and $f_2(t)$ are shown in the
     figure below. Sketch your result.

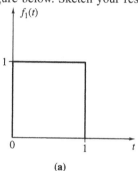 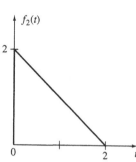

(a)   (b)

3.8 Rework Problem 3.5 using the time convolution theorem. Note how easy it is to find
     the frequency content of the output.

3.9 Prove the frequency convolution theorem by showing that the inverse Fourier
     transform of the right side of Eq. (3.3.17) is $f(t)g(t)$.

3.10 Rework Problem 3.7 using graphical convolution.

3.11 Graphically convolve $f_1(t)$ with $f_2(t)$, where $f_1(t)$ and $f_2(t)$ are given by

$$f_1(t) = 3[u(t+2) - u(t-2)]$$

and

$$f_2(t) = t[u(t) - u(t-2)] + (4-t)[u(t-2) - u(t-4)].$$

3.12 Graphically convolve $f_1(t)$ with $f_2(t)$, where $f_1(t)$ and $f_2(t)$ are shown in the figure
     below.

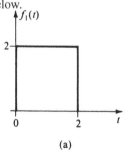 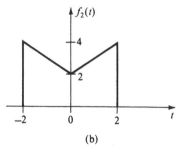

(a)   (b)

3.13 Find the output of an ideal LPF with $\omega_s = 377$ rad/sec, $A = 2.2$, and $t_d = 0$, for an input signal $r(t) = 10 + 2 \cos 72t - 14 \sin 300t + \sin 428t + \cos 1022t$.

3.14 Determine the output of an ideal BPF with $A = 1$, $t_d = 0$, $\omega_1 = 200$ rad/sec, and $\omega_2 = 500$ rad/sec, to the input signal $r(t)$ in Problem 3.13.

3.15 Repeat Problem 3.14 if $t_d = 10$ ms.

3.16 For an ideal HPF with $A = 3$, $\omega_L = 1000$ rad/sec, and $t_d = 0$, find the output for an input $r(t) = \sin 211t - \cos 306t - 8 \sin 830t + 7 \sin 1011t - \cos 3200t$.

3.17 Find the output of an ideal BSF with $A = 1$, $t_d = 0$, $\omega_1 = 500$ rad/sec, and $\omega_2 = 1000$ rad/sec, to the input $r(t)$ in Problem 3.13.

3.18 Find the output of the RC network in Fig. 3.3 for the input signal $r(t)$ in Problem 3.13 if $1/RC = 377$ rad/sec.

3.19 For a realizable BPF with the transfer function

$$H(\omega) = H_1(\omega + 250) + H_1(\omega - 250),$$

where

$$H_1(\omega) = \frac{1}{1 + j\omega RC}$$

and $RC = 1/377$, find the output time response for the input $r(t)$ in Problem 3.13.

3.20 The transfer function of the RC network in Fig. P3.20 is $H(\omega) = j\omega RC/(1 + j\omega RC)$. Does this network function as a LPF, BPF, BSF, or HPF? Which one? Find the output of this circuit to the input $r(t)$ in Problem 3.16 if $RC = 1/1000$.

3.21 The pulse shown in Fig. P3.21 is applied to the input of an ideal LPF with a step response given by Eq. (3.7.3). Write an expression for the filter output, $y(t)$.

3.22 By referring to Fig. 3.22 and a suitable table of Si(t) [2], sketch the system output in Problem 3.21 when $\tau = 100$ μsec for $f_s \gg 10$ kHz, $f_s = 10$ kHz, and $f_s \ll 10$ kHz. Note the relationship between bandwidth and output pulse shape.

3.23 Calculate the step response of the ideal LPF using the time convolution theorem, Eq. (3.5.5) and the Fourier transform of $u(t)$.

3.24 For a linear time-invariant system, it is possible to show that if

Fig. P3.20 Circuit for Problem 3.20

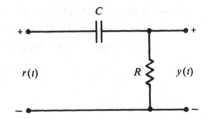

Fig. P3.21 Pulse for Problem
3.21

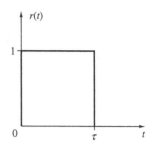

$$r_1(t) * h(t) = y_1(t),$$

then

$$\left[\frac{d}{dt}r_1(t)\right] * h(t) = \frac{d}{dt}y_1(t).$$

Using this result, demonstrate that if

$$r_2(t) * h(t) = y_2(t),$$

then

$$\left[\int_{-\infty}^{t} r_2(\tau)d\tau\right] * h(t) = \int_{-\infty}^{t} y_2(\tau)d\tau.$$

3.25 Use the result of Problem 3.24 to find the unit step response of the ideal LPF from $h_{\mathrm{LPF}}(t)$ in Eq. (3.5.6).

3.26 (a) Consider the time function $g_1(t)$ that has the Fourier transform $\mathcal{F}\{g_1(t)\} = G_1(\omega)$ shown in Fig. P3.26a. Can $g_1(t)$ be recovered from appropriately placed samples of $g_1(t)$? If so, what is the required sampling rate?

(b) Repeat part (a) for $G_2(\omega)$ in Fig. P3.26b.

3.27 The time function $g(t)$ with a Fourier transform $G(\omega)$ shown in Fig. P3.27 is sampled using shaped-top pulses by the infinite pulse train in Fig. 1.3.1. If $\tau = 0.1$ ms, write an expression for the Fourier transform of the sampled signal, $G_s(\omega)$. Can $g(t)$ be recovered undistorted from $g_s(t)$?

Fig. P3.26 Functions for
Problem 3.26

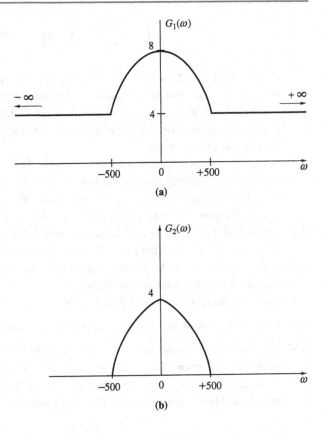

(a)

(b)

3.28 A voice signal, $m(t)$, which is bandlimited to 3200 Hz, multiplies the function cos $\omega_c t$, where $\omega_c = 2\pi(10,000)$ rad/sec. Assuming ideal sampling, specify a sampling rate such that $m(t) \cos \omega_c t$ can be uniquely recovered using an ideal BPF. What are the ideal BPF cutoff frequencies?

3.29 A time signal $g(t)$ is ideally sampled by the infinite train of impulses

Fig. P3.27 Fourier Transform
for Problem 3.27

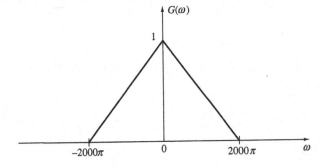

$$\delta_T\left(t - \frac{T}{2}\right) = \sum_{n=-\infty}^{\infty} \delta\left(t - nT - \frac{T}{2}\right).$$

Assuming that T satisfies the sampling theorem, can the original time function be reconstructed from these samples undistorted? Justify your answer.

3.30 Use Eq. (3.9.2) to find the energy in $y(t)$ in Eq. (3.3.11). Find an expression for the energy density spectrum of $y(t)$ and set up the integral for the total energy in $y(t)$ in terms of its energy density.

3.31 A signal $f(t)$ that is bandlimited to ω_m rad/sec has a Fourier transform denoted by $F(\omega)$. Find the energy density spectrum of $f(t) \cos \omega_c t$, where $\omega_c \gg \omega_m$.

3.32 Find the power density spectrum of $f(t) \cos \omega_c t$ if $f(t)$ is bandlimited to ω_m rad/sec and has the power density spectrum $S_f(\omega) = \lim_{\tau \to \infty}(1/2\tau)| F_\tau(\omega)|^2$. Let $\omega_c \gg \omega_m$.

3.33 The signal $A \cos \omega_c t$ is applied to the input of a system with impulse response $h(t) = e^{-t/\beta}u(t)$. Find the power density spectrum of the output.

3.34 If the signal $f(t) = Ae^{-\alpha t}u(t)$ is applied to the input of a differentiator, find the energy density spectrum of the output.

3.35 Find the energy in the signal $f(t) = \sqrt{t}e^{-\alpha t}u(t)$.

3.36 Finish the proof of Eq. (2.6.16) and show what happens if $\int_{-\infty}^{\infty} g(\lambda)d\lambda = 0$.

3.37 We would like to show that we can sample with flat-topped, finite width pulses with negligible distortion. The signal to be sampled is $g(t)$, and the flat-top pulses are generated by passing $g(t)$ through a sample-and-hold circuit. We assume that the sample-and-hold circuit behaves like a system with impulse response

$$h(t) = \begin{cases} 1, & \text{for } -\frac{\tau}{2} \le t \le \frac{\tau}{2} \\ 0, & \text{otherwise} \end{cases}$$

excited by the signal

$$r(t) = g(t)\delta_T(t) = \sum_{n=-\infty}^{\infty} g(nT)\delta(t - nT).$$

If the output of the sample-and-hold circuit is $g_s(t)$, show that

$$G_s(\omega) = \mathcal{F}\{g_s(t)\} = \omega_0\tau\frac{\sin(\omega\tau/2)}{\omega\tau/2}\sum_{n=-\infty}^{\infty}G(\omega - n\omega_0).$$

Sketch this result for some $G(\omega)$.

References

1. Kuo, F. F. 1962. *Network Analysis and Synthesis*. New York: Wiley.
2. Jahnke, E., and F. Emde. 1945. *Tables of Functions*. New York: Dover.

Random Variables and Stochastic Processes

4

4.1 Introduction

Randomness enters into communications problems through the classes of sources, such as speech, images, and music, that we wish to send from one location to another, as well as through the types of distortions, such as additive noise, phase jitter, and fading, that are inserted by real channels. Randomness is the essence of communications; it is that which makes communications both difficult and interesting. Everyone has an intuitive idea as to what "randomness" means, and in this chapter we develop different ways to characterize random phenomena that will allow us to account for randomness in both sources and channels in our communication system analyses and designs.

The treatment in this chapter is necessarily terse and incomplete. Every undergraduate engineer should have at least one course on the topics of probability and random variables, and this chapter does not purport to replace such a course. We simply survey the salient ideas and concepts of probability, random variables, and stochastic processes and define the requisite notation. The minimum mathematical mechanics needed to manipulate random variables and stochastic processes are also included, and after reading this material, the student will be prepared for probability and random processes wherever they are required.

As is typical of undergraduate communications books, theoretical details are suppressed to allow applications to be treated almost immediately. This is certainly justifiable, but the reader should be aware that rigorous proofs of many of the results in this chapter, as well as extensions, may require considerable mathematical maturity beyond our cursory treatment.

J. D. Gibson, *Fourier Transforms, Filtering, Probability and Random Processes*,
Synthesis Lectures on Communications, https://doi.org/10.1007/978-3-031-19580-8_4

4.2   Probability

A random experiment is an experiment whose outcome cannot be specified accurately in advance. Typical examples of a random experiment are the toss of a coin, the roll of a die, or the number of telephones in use at a particular time in a given telephone exchange. We denote the possible outcomes of a random experiment by $\zeta_i, i = 1, 2, \ldots, N$, and the set of all possible outcomes or the *sample space* by $S = \{\zeta_1, \zeta_2, \ldots, \zeta_N\}$. The number of outcomes (N) in the sample space may be finite, countably infinite, or uncountable. Finite sample spaces are the simplest to handle mathematically, and many of the examples familiar to the reader have a finite sample space. As a result, our initial discussion assumes that N is finite.

For a given random experiment, we may only be interested in certain subsets of the sample space or collections of outcomes. We define these subsets of the sample space to be *events*, and we call attention to two special events, the certain event S and the impossible event, which is the empty set $\emptyset$. That is, since S is the set of all possible outcomes of the random experiment, it occurs at every replication of the experiment, while the empty set or null set $\emptyset$ contains no elements, and hence can never occur.

We assign to each event $A \in S$ a nonnegative real number $P(A)$ called the *probability* of the event A. Every probability must satisfy three axioms:

Axiom 1: $P(A) \geq 0$

Axiom 2: $P(S) = 1$

Axiom 3: If $A \cap B = \emptyset$, then $P(A \cup B) = P(A) + P(B)$, where A and B are events.[1]

For simplicity of notation, it is common to write $A \cap B$ as AB and $A \cup B$ as $A + B$. We will use both notations as is convenient. If A^c denotes the complement of set A, it follows from Axioms 1–3 that

$$P(A^c) = 1 - P(A). \tag{4.2.1}$$

Further, using set operations and the axioms, it is possible to show that if $A \cap B \neq \emptyset$, then

$$P(A \cup B) = P(A) + P(B) - P(A \cap B). \tag{4.2.2}$$

To define probabilities for a countably infinite number of outcomes, we need the concept of a *field*. A class of sets that is closed under the set operations of complementation and finite unions and intersections is called a *field* $\mathcal{F}$. Thus a field $\mathcal{F}$ is a nonempty class of sets such that:

---

[1] The symbols "$\cap$" and "$\cup$" represent the set operations intersection and union, respectively. We shall not review set theory here.

1. If $A \in \mathcal{F}$, then $A^c \in \mathcal{F}$,
2. If $A \in \mathcal{F}$, and $B \in \mathcal{F}$, then $A + B \in \mathcal{F}$.

From (1) and (2) we can also show that if $A \in \mathcal{F}$ and $B \in \mathcal{F}$, then $A \cap B \in \mathcal{F}$ and $A - B \in \mathcal{F}$. Further, we have that $S \in \mathcal{F}$ and $\emptyset \in \mathcal{F}$. Thus it is evident that any finite number of set operations can be performed on the elements of $\mathcal{F}$ and still produce an element in $\mathcal{F}$.

A field is called a *Borel field* on S if it is closed under a countably infinite number of unions and intersections. Symbolically, a *Borel field*, denoted by $\mathcal{B}$, is a field defined on S such that if $A_1, A_2, \ldots, A_n, \ldots, \in \mathcal{B}$, then $\bigcup_{i=1}^{\infty} A_i \in \mathcal{B}$.

Using this definition, we can also show that if $A_1, A_2, \ldots, A_n, \ldots, \in \mathcal{B}$, then $\bigcap_{i=1}^{\infty} A_i \in \mathcal{B}$. Axioms 1 nad 2 remain valid for any event $A \in \mathcal{B}$ and we can extend Axiom 3 to a countably infinite sequence of mutually exclusive events (events that correspond to disjoint sets) in $\mathcal{B}$ as

Axiom 3'. For the mutually exclusive events $A_1, A_2, \ldots, A_n, \ldots, \in \mathcal{B}$,

$$P\left[\bigcup_{i=1}^{\infty} A_l\right] = \sum_{i=1}^{\infty} P(A_i).$$

We cannot proceed to define probabilities for an uncountably infinite number of outcomes without the mathematics of measure theory, which is well beyond the scope of the book. However, Axioms 1, 2, and 3' give us sufficient tools upon which to base our development.

An important concept in probability and in communications is the concept of independence. Two events A and B are said to be *independent* if and only if

$$P(A \cap B) = P(A)P(B). \tag{4.2.3}$$

Further, it often happens that we wish to calculate the probability of a particular event given the information that another event has occurred. Therefore, we define the conditional probability of B given A as $[P(A) \neq 0]$.

$$P(B|A) \triangleq \frac{P(A \cap B)}{P(A)}. \tag{4.2.4}$$

Note that if A and B are independent, then $P(B|A) = P(B)$; that is, knowing that A occurred does not change the probability of B. In calculating the probability of an event, it is often useful to employ what is called the *theorem of total probability*. Consider a finite or countably infinite collection of mutually exclusive $\left(A_i \cap A_j = \emptyset \text{ for all } i \neq j\right)$ and exhaustive $\left(\bigcup_i A_i = S\right)$ events denoted by $A_1, A_2, \ldots, A_n, \ldots$. Then the probability of an arbitrary event B is given by

$$P(B) = \sum_i P(A_i \cap B) = \sum_i P(A_i)P(B|A_i). \tag{4.2.5}$$

Example 4.2.1 To illustrate the preceding concepts, we consider the random experiment of observing whether two telephones in a particular office are busy (in use) or not busy. The outcomes of the random experiment are thus:

$$\zeta_1 = \text{neither telephone is busy}$$
$$\zeta_2 = \text{only telephone number 1 is busy}$$
$$\zeta_3 = \text{only telephone number 2 is busy}$$
$$\zeta_4 = \text{both telephones are busy.}$$

We assign probabilities to these outcomes according to $P(\zeta_1) = 0.2$, $P(\zeta_2) = 0.4$, $P(\zeta_3) = 0.3$, and $P(\zeta_4) = 0.1$. .. We can also define the sets (events) $A = \{\zeta_1\}$, $B = \{\zeta_2\}$, $C = \{\zeta_3\}$, and $D = \{\zeta_4\}$, so $P(A) = 0.2$, $P(B) = 0.4$, $P(C) = 0.3$, and $P(D) = 0.1$.

(a) What is the probability of the event that one or more telephones are busy?

$$P(\text{one or more busy}) = P(B \cup C \cup D),$$

which we could evaluate by repeated application of Eq. (4.2.2). However, we note that $A = (B \cup C \cup D)^c$ so we can use Eq. (4.2.1) to obtain

$$P(\text{one or more busy}) = 1 - P(A) = 0.8.$$

(b) What is the probability that telephone number 1 is in use?
$P(\text{telephone 1 busy}) = P(B \cup D) = P(B) + P(D) = 0.5$, since $B \cap D = \emptyset$
(c) Let E be the event that telephone number 1 is busy and F be the event that telephone number 2 is busy. Are E and F independent?
We need to check if Eq. (4.2.3) is satisfied. From (b) we have $P(E) = 0.5$, and in a similar fashion we obtain $P(F) = 0.4$. Hence

$$P(E)P(F) = 0.2.$$

Since $D = E \cap F$,

$$P(E \cap F) = P(D) = 0.1.$$

Thus $P(EF) \neq P(E)P(F)$ and E and F are not independent.
(d) Referring to event E in part (c), find $P(D|E)$.
From Eq. (4.2.4),

$$P(D|E) = \frac{P(DE)}{P(E)} = \frac{P(E|D)P(D)}{P(E)}.$$

But $P(E|D) = 1$, so

$$P(D|E) = \frac{P(D)}{P(E)} = 0.2.$$

Note that $P(D|E) \neq P(D)$; hence knowing that E has occurred changes the probability that D will occur.

4.3   Probability Density and Distribution Functions

The outcomes of a random experiment need not be real numbers. However, for purposes of manipulation, it would be much more convenient if they were. Hence we are led to define the concept of a random variable.

Definition 4.3.1 A real random variable $X(\zeta_i)$ is a function that assigns a real number, called the value of the random variable, to each possible outcome of a random experiment.

Therefore, for Example 4.2.1 we could define the random variable $X(\zeta_1) = 0$, $X(\zeta_2) = 1$, $X(\zeta_3) = 2$, and $X(\zeta_4) = 3$. Note that for this random variable, values are assigned only for discrete points and the values in between are not allowable. Such a random variable is said to be discrete. If a random variable can take on a continuum of values in some range, the random variable is said to be continuous. It is also possible to define mixed random variables that have both discrete and continuous components.

Of course, to work with random variables, we need a probabilistic description of their behavior. This is available from the probabilities assigned to the individual outcomes of the random experiment. That is, for discrete random variables, we can write

$$P[X(\zeta_i) = x_i] = P[\zeta_i \in S : X(\zeta_i) = x_i], \tag{4.3.1}$$

which simply means that the probability of the value of the random variable x_i is just the probability of the outcome ζ_i associated with x_i through $X(\zeta_i)$. The right-hand side of Eq. (4.3.1) is read as "the probability of $\zeta_i \in S$ such that $X(\zeta_i) = x_i$." For continuous random variables, the probability of $X(\zeta) = x$ is zero, but the probability of $X(\zeta)$ taking on a range of values is nonzero. Hence we have

$$P[X(\zeta_i) \leq x_i] = P[\zeta_i \in S : X(\zeta_i) \leq x_i] \tag{4.3.2}$$

for continuous random variables. It is common to drop the explicit dependence on the experimental outcome from the left side of Eq. (4.3.2) and write for some outcome ζ,

$$P[X \leq x] = P[X(\zeta) \leq x]. \tag{4.3.3}$$

We can just as well talk about other forms of intervals than $X \leq x$, but this form has a special meaning.

Definition 4.3.2 The cumulative distribution function (CDF), or simply distribution function of a random variable X, is given by.

$$F_X(x) = P[X(\zeta) \leq x] = P[X \leq x],$$ (4.3.4)

for $-\infty < x < \infty$.

A distribution function has the properties

$$0 \leq F_X(x) \leq 1$$ (4.3.5a)

$$F_X(x_1) \leq F_X(x_2) \text{ for } x_1 < x_2$$ (4.3.5b)

$$F_X(-\infty) \triangleq \lim_{x \to -\infty} F_X(x) = 0$$ (4.3.5c)

$$F_X(\infty) \triangleq \lim_{x \to \infty} F_X(x) = 1$$ (4.3.5d)

$$\lim_{\substack{\varepsilon \to 0 \\ \varepsilon > 0}} F_X(x + \varepsilon) \triangleq F_X(x^+) = F_X(x).$$ (4.3.5e)

The reader should note that Eq. (4.3.5e) indicates that the distribution function is continuous from the right. Some authors prefer to define the cumulative distribution function in terms of $P[X < x]$. that is, strict inequality, in which case the distribution function is continuous from the left. As long as the specific definition being used is clear, the subsequent details will be consistent. We use Eq. (4.3.4) as our definition throughout the book. Since $P[X \leq x] + P[X < x] = 1$, then

$$P[X > x] = 1 - F_X(x).$$ (4.3.6)

Further, we can write

$$P[x_1 < X \leq x_2] = P[X \leq x_2] - P[X \leq x_1] = F_X(x_2) - F_X(x_1).$$ (4.3.7)

If we let

$$F_X(x^-) = \lim_{\substack{\varepsilon \to 0 \\ \varepsilon > 0}} F_X(x - \varepsilon),$$

then

$$P[X = x] = P[X \le x] - P[X < x] = F_X(x) - F_X(x^-). \qquad (4.3.8)$$

For a continuous random variable, $F_X(x^-) = F_X(x)$, so $P[X = x] = 0$. However, for a discrete distribution function, there will be several points where $F_X(x)$ and $F_X(x^-)$ differ, and hence $F_X(x)$ will contain "jumps" or discontinuities at these points.

It is convenient to define another function, called the *probability density function* (pdf), denoted here by $f_X(x)$. For a discrete random variable, the probability density function is given by

$$f_X(x) = \begin{cases} P[X = x_i], & \text{for } x = x_i, i = 1, 2, \ldots \\ 0, & \text{for all other } x, \end{cases} \qquad (4.3.9)$$

where $P[X = x_i]$ is computed from Eq. (4.3.8). Thus, for a discrete random variable, $f_X(x)$ is nonzero only at the points of discontinuity of $F_X(x)$. Now, if we write $F_X(x)$ in terms of the unit step function, that is,

$$F_X(x) = P[X \le x] = \sum_{x_i} P[X = x_i] u(x - x_i), \qquad (4.3.10)$$

we can obtain an expression for $f_X(x)$ by formally taking the derivative of Eq. (4.3.10) as

$$f_X(x) \triangleq \frac{d}{dx} F_X(x) = \sum_x P[X = x_i] \delta(x - x_i) \qquad (4.3.11)$$

for $-\infty < x < \infty$. We see by inspection that Eqs. (4.3.9) and (4.3.11) agree. Note that it is not strictly mathematically correct to use impulses in Eq. (4.3.11), since the impulses actually have infinite amplitude. However, this is a common methodology and it allows us to treat continuous and discrete random variables in a unified fashion. A more careful approach would be differentiation of the cumulative distribution function to obtain the pdf in the continuous case and differencing as in Eq. (4.3.8) in the discrete case.

For a continuous random variable X with distribution function $F_X(x)$, we have that

$$f_X(x) = \frac{d}{dx} F_X(x) \ge 0 \qquad (4.3.12)$$

for all x, and further,

$$F_X(x) = \int_{-\infty}^{x} f_X(\lambda) d\lambda. \qquad (4.3.13)$$

Using Eq. (4.3.13) with Eqs. (4.3.5c) and (4.3.5d), we find that

$$\int_{-\infty}^{\infty} f_X(\lambda) d\lambda = 1. \qquad (4.3.14)$$

Similarly, we have for a discrete random variable that

$$\int_{-\infty}^{\infty} f_X(x)dx = \sum_{\text{all } i} P[X = x_i] = 1. \tag{4.3.15}$$

Example 4.3.1 We are given the distribution function of a discrete random variable Y,

$$F_Y(y) = \begin{cases} 0, & y < -1 \\ \frac{1}{8}, & -1 \le y < 0 \\ \frac{3}{8}, & 0 \le y < 2 \\ \frac{3}{4}, & 2 \le y < 3 \\ 1, & y \ge 3. \end{cases}$$

(a) Sketch $F_Y(y)$.

A sketch of $F_Y(y)$ is shown in Fig. 4.1. The heavy dots at the step points indicate that $F_Y(y)$ is continuous from the right.

(b) Write an expression for $F_Y(y)$.

In terms of unit step functions, we have

$$F_Y(y) = \frac{1}{8}u(y+1) + \frac{1}{4}u(y) + \frac{3}{8}u(y-2) + \frac{1}{4}u(y-3).$$

(c) Write an expression for and sketch the probability density function of Y.

Using Eq. (4.3.11), we can write

$$f_Y(y) = \sum_{y_i} P[Y = y_i]\delta(y - y_i)$$

Fig. 4.1 Distribution function $F_Y(y)$ for Example 4.3.1

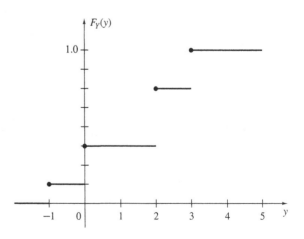

Fig. 4.2 Probability density function $f_Y(y)$ for Example 4.3.1

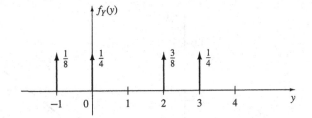

$$= \frac{1}{8}\delta(y+1) + \frac{1}{4}\delta(y) + \frac{3}{8}\delta(y-2) + \frac{1}{4}\delta(y-3).$$

A sketch of this pdf is shown in Fig. 4.2.

(d) Find $P[Y \le 0]$, $P[Y \le 1]$, and $P[Y > 2]$.

$$P[Y \le 0] = F_Y(0) = \frac{1}{8} + \frac{1}{4} = \frac{3}{8}$$

$$P[Y \le 1] = F_Y(1) = \frac{1}{8} + \frac{1}{4} = \frac{3}{8}$$

$$P[Y > 2] = 1 - P[Y \le 2] = 1 - F_Y(2)$$

$$= 1 - \left[\frac{1}{8} + \frac{1}{4} + \frac{3}{8}\right] = \frac{1}{4}.$$

Example 4.3.2 The distribution function of a continuous random variable Z is given by.

$$F_Z(z) = \begin{cases} 0, & z < 2 \\ \frac{z-2}{4}, & 2 \le z < 6 \\ 1, & z \ge 6. \end{cases}$$

(a) Sketch $F_Z(z)$. See Fig. 4.3.

(b) Find an expression for and sketch the pdf of Z.

$$f_Z(z) = \begin{cases} 0, & z < 2 \\ \frac{1}{4}, & 2 \le z < 6 \\ 0, & z \ge 6. \end{cases}$$

See the sketch in Fig. 4.4.

(c) Find $P[Z \le 3]$ and $P[4 < Z \le 6]$.

$$P[Z \le 3] = F_Z(3) = \frac{1}{4}$$

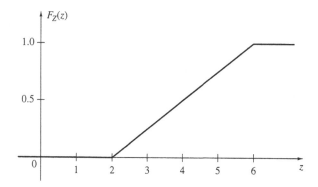

Fig. 4.3 Sketch of $F_Z(z)$ in Example 4.3.2

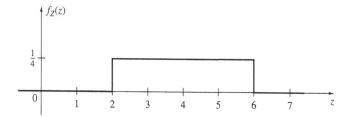

Fig. 4.4 Sketch of the pdf of Z for Example 4.3.2

$$P[4 < Z \leq 6] = F_Z(6) - F_Z(4) = 1 - \frac{1}{2} = \frac{1}{2}.$$

It is possible to extend the concepts of distribution functions and density functions to more than one random variable in a straightforward manner. For example, for the case of two random variables X and Y, we can define their joint distribution, also called the bivariate distribution of X and Y, as

$$F_{XY}(x, y) = P[X \leq x, Y \leq y], \tag{4.3.16}$$

where properties analogous to the scalar case hold. Assuming that the joint probability density function of X and Y exists, we can also write Eq. (4.3.16) as

$$F_{XY}(x, y) = \int_{-\infty}^{y} \int_{-\infty}^{x} f_{XY}(\lambda, \gamma) d\lambda \, d\gamma. \tag{4.3.17}$$

Furthermore, given the joint distribution function $F_{XY}(x, y)$, we can obtain the probability density function from

$$f_{XY}(x, y) = \frac{\partial^2}{\partial x \partial y} F_{XY}(x, y). \tag{4.3.18}$$

Of course, it is often the case that we are given a bivariate distribution and we need the univariate distribution function of one of the random variables. The univariate distributions are also called the marginal distributions and can be obtained straightforwardly as

$$F_X(x) = F_{XY}(x, \infty) = \int_{-\infty}^{x} \int_{-\infty}^{\infty} f_{XY}(\lambda, \gamma) d\gamma \, d\lambda$$

$$= \int_{-\infty}^{x} f_X(\lambda) d\lambda. \tag{4.3.19}$$

Similarly, the marginal density is given by

$$f_X(x) = \int_{-\infty}^{\infty} f_{XY}(x, \gamma) d\gamma. \tag{4.3.20}$$

For purely discrete random variables, the integrals in Eqs. (4.3.17), (4.3.19), and (4.3.20) are replaced by summations.

There is no difficulty in extending joint distribution and density functions to the multivariate (more than two variables) case.

Example 4.3.3 Given the joint probability density function

$$f_{XY}(x, y) = \begin{cases} Ke^{-(x+y)}, & \text{for } x \geq 0, \quad y \geq 0 \\ 0, & \text{elsewhere.} \end{cases}$$

(a) Find K such that $f_{XY}(x, y)$ is a valid pdf. We need

$$\int_{-\infty}^{\infty} \int_{-\infty}^{\infty} f_{XY}(x, y) dx \, dy = 1,$$

so we find that

$$K \int_{0}^{\infty} \int_{0}^{\infty} e^{-(x+y)} dx \, dy = K = 1.$$

(b) Calculate $F_{XY}(x, y)$.

$$F_{XY}(x, y) = \int_{-\infty}^{y} \int_{-\infty}^{x} f_{XY}(\lambda, \gamma) d\lambda \, d\gamma$$

$$= \int_0^y \int_0^x e^{-(\lambda+\gamma)} d\lambda \, d\gamma = \left(1 - e^{-x}\right)\left(1 - e^{-y}\right)$$

for $x \geq 0$, $y \geq 0$, and $F_{XY}(x, y) = 0$ otherwise.

(c) Find the marginal distribution and density functions of Y. Directly, the distribution function of Y is

$$F_Y(y) = F_{XY}(\infty, y) = \left.\left(1 - e^{-x}\right)\right|_{x=\infty} \left(1 - e^{-y}\right)$$
$$= 1 - e^{-y} \text{ for } y \geq 0$$

$$F_Y(y) = 0, \quad \text{for } y < 0.$$

We can find the pdf of Y by differentiating $F_Y(y)$ or by integrating over all x in $f_{XY}(x, y)$. Thus

$$f_Y(y) = \frac{d}{dy} F_Y(y) = \begin{cases} e^{-y}, & y \geq 0 \\ 0, & y < 0. \end{cases}$$

(d) Find $P[X \leq 2, Y \leq 5]$.

By substitution in $F_{XY}(x, y)$,

$$P[X \leq 2, Y \leq 5] = F_{XY}(2, 5) = \left(1 - e^{-2}\right)\left(1 - e^{-5}\right).$$

(e) Find $P[X \leq 2, Y > 1]$.

The simplest way to evaluate this probability is to return to the joint density and write

$$P[X \leq 2, Y > 1] = \int_1^\infty \int_0^2 e^{-(x+y)} dx \, dy = \left(1 - e^{-2}\right) e^{-1}.$$

Analogous to independent events in a random experiment, we can have independent random variables.

Definition 4.3.3 Two random variables X and Y are said to be independent if and only if.

$$F_{XY}(x, y) = F_X(x) \, F_Y(y). \tag{4.3.21}$$

Of course, this condition is equivalent to

$$f_{XY}(x, y) = f_X(x) \, f_Y(y). \tag{4.3.22}$$

Therefore, two random variables are independent if and only if their joint distribution (or density) factors into the product of their marginal distributions (densities).

Example 4.3.4 Are the random variables X and Y in Example 4.3.3 independent? Examining the pdfs, we know from Example 4.3.3 that.

$$f_{XY}(x, y) = \begin{cases} e^{-(x+y)}, & x \geq 0, \quad y \geq 0 \\ 0, & \text{otherwise} \end{cases}$$

and

$$f_Y(y) = \begin{cases} e^{-y}, & y \geq 0 \\ 0, & y < 0. \end{cases}$$

Straightforwardly, we find that

$$f_X(x) = \begin{cases} e^{-x}, & x \geq 0 \\ 0, & x < 0. \end{cases}$$

Hence

$$f_{XY}(x, y) = f_X(x) \, f_Y(y)$$

and X and Y are independent. It is simple to demonstrate the same result using distribution functions.

Just as we defined the conditional probability of event B occurring given that event A has occurred in Eq. (4.2.4), we can define conditional distribution and density functions. Given two random variables X and Y, we can define the conditional distribution function of X given that $Y \leq y$ as

$$F_{X|Y}(x|y) = \frac{F_{XY}(x, y)}{F_Y(y)}, \tag{4.3.23}$$

where $F_Y(y) \neq 0$. If the conditioning is that $Y = y$ (rather than $Y \leq y$), we have

$$F_{X|Y}(x|Y = y) = \frac{\int_{-\infty}^{x} f_{XY}(\lambda, y) d\lambda}{f_Y(y)}. \tag{4.3.24}$$

The distribution functions in Eqs. (4.3.23) and (4.3.24) are both valid univariate distribution functions, and hence satisfy the properties in Eq. (4.3.5).

From Eq. (4.3.24) we can also find the conditional probability density of X given $Y = y$ by differentiation as

$$f_{X|Y}(x|y) = \frac{f_{XY}(x, y)}{f_Y(y)}. \tag{4.3.25}$$

Similarly, we can obtain

$$f_{Y|X}(y|x) = \frac{f_{XY}(x, y)}{f_X(x)}, \tag{4.3.26}$$

so we can combine Eqs. (4.3.25) and (4.3.26) to get a form of Bayes' rule,

$$f_{Y|X}(y|x) = \frac{f_{X|Y}(x|y) f_Y(y)}{f_X(x)}. \tag{4.3.27}$$

Example 4.3.5 We are given the joint pdf of X and Y,

$$f_{XY}(x, y) = \begin{cases} x + y, \ 0 \le x \le 1, \quad 0 \le y \le 1 \\ 0, \qquad \text{otherwise.} \end{cases}$$

(a) Find the marginal densities of X and Y.

$$f_X(x) = \begin{cases} \int_0^1 (x + y)dy = \left[xy + \frac{y^2}{2}\right]\Big|_0^1 = x + \frac{1}{2}, \ \text{for} \ 0 \le x \le 1 \\ 0, \qquad\qquad\qquad\qquad\qquad\qquad\qquad \text{otherwise} \end{cases}$$

Similarly, we can show that

$$f_Y(y) = \begin{cases} y + \frac{1}{2}, \ 0 \le y \le 1 \\ 0, \qquad \text{otherwise.} \end{cases}$$

(b) Find $f_{X|Y}(x|y)$.
    From Eq. (4.3.25),

$$f_{X|Y}(x|y) = \frac{f_{XY}(x, y)}{f_Y(y)},$$

so

$$f_{X|Y}(x|y) = \begin{cases} \frac{x+y}{y+\frac{1}{2}}, \ 0 \le x \le 1 \\ 0, \qquad \text{otherwise.} \end{cases}$$

(c) Calculate $F_{X|Y}(x|Y = y)$.
    From part (b),

$$F_{X|Y}(x|Y = y) = \int_0^x \frac{\lambda + y}{y + \frac{1}{2}} d\lambda = \frac{1}{y + \frac{1}{2}}\left[\frac{\lambda^2}{2} + \lambda y\right]\Big|_0^x = \frac{x^2/2 + xy}{y + \frac{1}{2}}$$

for $0 \le x \le 1$, and

$$F_{X|Y}(x|Y = y) = \begin{cases} 0, & x < 0 \\ 1, & x \ge 1. \end{cases}$$

(d) Calculate $F_{X|Y}(x|y)$.

From Eq. (4.3.23), we see that we need $F_Y(y)$, so

$$F_Y(y) = \int_0^y \left(\gamma + \frac{1}{2} \right) d\gamma = \left[\frac{\gamma^2}{2} + \frac{\gamma}{2} \right]\Big|_0^y$$

$$= \frac{y^2}{2} + \frac{y}{2} = \frac{y}{2}(y + 1), \quad 0 \le y \le 1$$

$$= 0, \quad y < 0$$

$$= 1, \quad y \ge 1.$$

Also,

$$F_{XY}(x, y) = \int_0^y \int_0^x (\lambda + \gamma) d\gamma \, d\lambda$$

$$= \frac{xy}{2}(x + y), \quad 0 \le x \le 1, \quad 0 \le y \le 1$$

$$F_{XY}(x, y) = 0, \quad x < 0, \quad y < 0$$

$$= 1, \quad x \ge 1, \quad y \ge 1.$$

$$F_{X|Y}(x|y) = \frac{F_{XY}(x, y)}{F_Y(y)}$$

$$= \frac{x(x + y)}{y + 1}, \quad 0 \le x \le 1$$

$$F_{X|Y}(x|y) = 0, \quad x < 0$$

$$= 1, \quad x \ge 1$$

Note that $F_{X|Y}(x|y) \ne F_{X|Y}(x|Y = y)$.

4.4   Mean, Variance, and Correlation

The probability density function or the cumulative distribution function describes in detail the behavior of the particular random variable being considered. However, there are many applications where we need to work with numbers that are representative of the particular distribution function of interest without carrying along the entire pdf or CDF. Perhaps the single most-used parameter of the distribution of a random variable is the expected value, also called the expectation or mean value.

Definition 4.4.1 The expected value of a random variable X is given by.

$$E\{X\} \triangleq \int_{-\infty}^{\infty} x f_X(x) dx, \tag{4.4.1}$$

where $f_X(X)$ is the pdf of X. For a discrete random variable X, Eq. (4.4.1) becomes

$$E\{X\} = \sum_i x_i P[X = x_i]. \tag{4.4.2}$$

It is common to represent $E\{X\}$ by the symbol μ_X.

It is also possible to write the expectation of a function of the random variable X, say $g(X)$, in terms of an integral over the pdf of X. This result, sometimes called the fundamental theorem of expectation, is given by

$$E\{g(X)\} = \int_{-\infty}^{\infty} g(x) f_X(x) dx. \tag{4.4.3}$$

We do not offer a proof of Eq. (4.4.3), and indeed, it is not trivial to prove in all generality. Of course, in Eq. (4.4.3) we could let $g(x) = x$, which would yield Eq. (4.4.1), or $g(x)$ can be any function of x. Of particular importance is when $g(x) = x^2$, so Eq. (4.4.3) becomes

$$E\{X^2\} = \int_{-\infty}^{\infty} x^2 f_X(x) dx, \tag{4.4.4}$$

which is called the second moment of X, or the second moment about zero.

Next to the expectation, perhaps the second most widely used parameter of the probability distribution of a random variable is the variance. The variance of a random variable X is defined as

$$\text{var}\{X\} \triangleq E\{[X - \mu_X]^2\} = E\{X^2\} - \mu_X^2. \tag{4.4.5}$$

The variance of X is also the second moment about the mean of X. Thus, while μ_X locates the mean or average value of the random variable, the variance is an indicator of the dispersion about the mean. The variance is often denoted by σ_X^2, and the positive square root of the variance is called the *standard deviation* and denoted by σ_X.

Example 4.4.1 Given a random variable X with pdf

$$f_X(x) = \begin{cases} e^{-x}, & x \geq 0 \\ 0, & x < 0. \end{cases}$$

(a) Find $E\{X\}$.

  By Eq. (4.4.1),

$$E\{X\} = \int_0^\infty xe^{-x}dx = \left[xe^{-x} - e^{-x}\right]\Big|_0^\infty = 1.$$

(b) Find $E\{X^2\}$.

  From Eq. (4.4.4),

$$E\{X^2\} = \int_0^\infty x^2e^{-x}dx = 2.$$

(c) What is the variance of X?

  By Eq. (4.4.5),

$$\text{var}(X) = E\{X^2\} - \mu_X^2 = 2 - 1 = 1.$$

(d) Given $g(X) = ax + b$, what is $E\{g(X)\}$?

  Using Eq. (4.4.3), we obtain

$$E\{g(X)\} = E\{ax + b\} = \int_0^\infty [ax + b]e^{-x}dx$$

$$= a \int_0^\infty xe^{-x}dx + b \int_0^\infty e^{-x}dx = a\mu_X + b = a + b.$$

It is important to note that the integral defining the expected value in Eq. (4.4.1) may not exist, in which case the mean does not exist. The most common example of this phenomenon is the Cauchy distribution, defined as

$$f_X(x) = \frac{1}{\pi(1 + x^2)}, \quad -\infty < x < \infty. \tag{4.4.6}$$

The demonstration that $E\{X\}$ does not exist for Eq. (4.4.6) is left as a problem.

An extremely convenient function for solving various problems involving random variables is obtained by letting $g(X) = e^{j\omega X}$ in Eq. (4.4.3).

Definition 4.4.2 The characteristic function of a random variable X is defined by.

$$\Phi_X(\omega) = E\left\{e^{j\omega X}\right\}. \tag{4.4.7}$$

Note that the characteristic function of X is simply the Fourier transform of the pdf of X with ω replaced by $-\omega$. Since the Fourier transform is unique, the characteristic function is also unique, and hence conclusions reached using characteristic functions can be uniquely translated into conclusions concerning pdfs. For example, given two random variables X and Y, if $\Phi_X(\omega) \equiv \Phi_Y(\omega)$, then $F_X(x) \equiv F_Y(y)$. [Since we are assuming that pdfs exist for our work, we can also conclude that $f_X(x) \equiv f_Y(y)$.] Also of great importance is the fact that if we let $E\{X^n\} = m_n$, then

$$\left. \frac{d^n}{d\omega^n} \Phi_X(\omega) \right|_{\omega=0} = j^n m_n. \tag{4.4.8}$$

Therefore, moments of a random variable X are easily obtainable from its characteristic function.

Example 4.4.2 Given the pdf

$$f_X(x) = \begin{cases} e^{-x}, & x \geq 0 \\ 0, & x < 0. \end{cases}$$

(a) Find $\Phi_X(\omega)$

$$\Phi_X(\omega) = \int_0^\infty e^{j\omega x} e^{-x} dx = \int_0^\infty e^{-(1-j\omega)x} dx$$

$$= \left. \frac{-1}{1-j\omega} e^{-(1-j\omega)x} \right|_0^\infty = \frac{1}{1-j\omega}.$$

(b) Calculate $E\{X\}$ and $E\{X^2\}$.
By Eq. (4.4.8),

$$E\{X\} = m_1 = \left. \frac{1}{j} \frac{d}{d\omega} \Phi_X(\omega) \right|_{\omega=0} = \left. \frac{1}{j} \left\{ \frac{j}{(1-j\omega)^2} \right\} \right|_{\omega=0} = 1.$$

Also by Eq. (4.4.8),

$$E\{X^2\} = m_2 = \left. \frac{1}{j^2} \frac{d^2}{d\omega^2} \Phi_X(\omega) \right|_{\omega=0} = \left. (-1) \left\{ \frac{-j^2(1-j\omega)(-j)}{(1-j\omega)^4} \right\} \right|_{\omega=0} = 2.$$

Both results check with Example 4.3.1.

We can also consider moments of joint distributions. Given two random variables X and Y with joint pdf $f_{XY}(x, y)$, we can define

$$E\left\{X^n Y^k\right\} = \int_{-\infty}^{\infty} \int_{-\infty}^{\infty} x^n y^k f_{XY}(x, y) dx\, dy. \tag{4.4.9}$$

In fact, we have the more general result, analogous to the univariate case in Eq. (4.4.3), that

$$E\{g(X, Y)\} = \int_{-\infty}^{\infty} \int_{-\infty}^{\infty} g(x, y) f_{XY}(x, y) dx\, dy. \tag{4.4.10}$$

We can also define bivariate characteristic functions as

$$\Phi_{XY}(\omega_1, \omega_2) = \int_{-\infty}^{\infty} \int_{-\infty}^{\infty} e^{j\omega_1 x + j\omega_2 y} f_{XY}(x, y) dx\, dy. \tag{4.4.11}$$

Letting $E\left\{X^n Y^k\right\} = m_{nk}$, we also have that

$$\left. \frac{\partial^k \partial^n \Phi(\omega_1, \omega_2)}{\partial \omega_2^k \partial \omega_1^n} \right|_{\omega_1 = \omega_2 = 0} = j^{(n+k)} m_{nk}. \tag{4.4.12}$$

Equations (4.4.9)–(4.4.11) are extended to more than two random variables in the obvious way.

Two random variables X and Y are said to be *uncorrelated* if

$$E\{XY\} = E\{X\}E\{Y\} \tag{4.4.13}$$

and orthogonal if

$$E\{XY\} = 0. \tag{4.4.14}$$

The *covariance* of X and Y is given by

$$\mathrm{cov}(X, Y) = E\{[X - \mu_X][Y - \mu_Y]\} = E\{XY\} - \mu_X \mu_Y, \tag{4.4.15}$$

from which we define the *correlation coefficient*

$$\rho_{XY} = \frac{\mathrm{cov}(X, Y)}{\sqrt{\mathrm{var}(X)\mathrm{var}(Y)}} = \frac{E\{XY\} - \mu_X \mu_Y}{\left[(E\{X^2\} - \mu_X^2)(E\{Y^2\} - \mu_Y^2)\right]^{1/2}}. \tag{4.4.16}$$

We note that

$$-1 \le \rho_{XY} \le 1. \tag{4.4.17}$$

Example 4.4.3 Let us reconsider the joint pdf of X and Y from Example 4.2.5,

$$f_{XY}(x, y) = \begin{cases} x + y, & 0 \le x \le 1, \quad 0 \le y \le 1 \\ 0, & \text{otherwise.} \end{cases}$$

(a) Find $E\{XY\}$.

From Eq. (4.4.9),

$$E\{XY\} = \int_0^1 \int_0^1 xy(x + y)dx\,dy = \int_0^1 \int_0^1 (x^2 y + xy^2)dx\,dy$$

$$= \int_0^1 \left[\frac{x^3 y}{3} + \frac{x^2 y^2}{2} \right]\Big|_0^1 dy = \int_0^1 \left[\frac{y}{3} + \frac{y^2}{2} \right] dy = \left[\frac{y^2}{6} + \frac{y^3}{6} \right]\Big|_0^1 = \frac{1}{3}.$$

(b) Calculate $\mathrm{cov}(X, Y)$ and ρ_{XY}

By Eq. (4.4.15), we need μ_X and μ_Y, so using the marginal pdfs from Example 4.3.5

$$\mu_X = \int_0^1 x\left(x + \frac{1}{2} \right)dx = \left[\frac{x^3}{3} + \frac{x^2}{4} \right]\Big|_0^1 = \frac{7}{12}$$

and $\mu_Y = \frac{7}{12}$. Thus

$$\mathrm{cov}(X, Y) = E\{XY\} - \mu_X \mu_Y = \frac{1}{3} - \left(\frac{7}{12} \right)^2 = \frac{-1}{144}.$$

Now

$$E\{X^2\} = \int_0^1 x^2\left(x + \frac{1}{2} \right)dx = \left[\frac{x^4}{4} + \frac{x^3}{6} \right]\Big|_0^1 = \frac{5}{12}.$$

and

$$E\{Y^2\} = \frac{5}{12},$$

so

$$\rho_{XY} = \frac{-\frac{1}{144}}{\left[\left(\frac{5}{12} - \frac{49}{144} \right)^2 \right]^{1/2}} = -\frac{1}{11}.$$

(c) Are X and Y uncorrelated? Orthogonal?

Since $E\{XY\} \ne 0$, they are not orthogonal, and further, $E\{XY\} = \frac{1}{3} \ne \mu_X \mu_Y = \frac{49}{144}$, so X and Y, uncorrelated.

(d) Find the joint characterisic function of X and Y.

From Eq. (4.4.11),

$$\Phi_{XY}(\omega_1, \omega_2) = \int_0^1 \int_0^1 e^{j\omega_1 x + j\omega_2 y}(x + y)\, dx\, dy$$

$$= \frac{j}{\omega_2 \omega_1^2}\left[1 - e^{j\omega_2}\right]\{e^{j\omega_1} - 1 - j\omega_1 e^{j\omega_1}\}$$

$$+ \frac{j}{\omega_1 \omega_2^2}\left[1 - e^{j\omega_1}\right]\{e^{j\omega_2} - 1 - j\omega_2 e^{j\omega_2}\}.$$

Note the Eq. (4.4.12) is not a good way to obtain joint moments of X and Y for this example.

Finally, we simply note that we can define the mean and variance for conditional distributions straightforwardly as

$$E\{X|Y = y\} = \int_{-\infty}^{\infty} x f_{X|Y}(x|y)dx \triangleq \mu_{X|Y} \qquad (4.4.18)$$

and

$$\text{var}\{X|Y = y\} = E\{X^2|Y = y\} - \mu_{X|Y}^2. \qquad (4.4.19)$$

The conditional mean plays a major role in the theory of minimum mean-squared error estimation. We will not be able to pursue this connection here.

4.5 Transformations of Random Variables

In electrical engineering and communications, it is often the case that we are given some random variable X with a known pdf $f_X(x)$, and we wish to obtain the pdf of a random variable Y related to X by some function $y = g(x)$. For discrete random variables, the determination of $f_Y(y)$ is quite straightforward. Given a discrete random variable X with a pdf specified by $f_X(x_i) = P[X = x_f]$ and a one-to-one transformation $y = g(x)$, we have

$$f_Y(y_i) = P[Y = y_i = g(x_j)] = P[X = x_j]. \qquad (4.5.1)$$

For continuous random variables, we can calculate $f_Y(y)$ by first solving $y = g(x)$ for all of its roots, denoted $x_i = g^{-1}(y_i)$, $i = 1, 2, \dots$. Then we can express $f_Y(y)$ in terms of $f_X(x)$ as

$$f_Y(y) = \sum_i \frac{f_X(x)}{|(d/dx)g(x)|}\Big|_{x=x_i} \qquad (4.5.2)$$

$$= \sum_i \left[f_x(x) \left| \frac{dx}{dy} \right| \right] \Bigg|_{x=x_i}. \tag{4.5.3}$$

Example 4.5.1 We are given a random variable X with pdf

$$f_x(x) = \begin{cases} e^{-x}, & x \geq 0 \\ 0, & x < 0 \end{cases}$$

and we wish to find the pdf of a random variable Y related to X by $Y = aX + b$.

Solving $y = g(x)$ for x, we find $x = (y - b)/a$, so from Eq. (4.5.2)

$$f_y(y) = \frac{e^{-x} u(x)}{|(d/dx)(ax + b)|} \Bigg|_{x=(y-b)/a} = \frac{1}{|a|} e^{-((y-b)/a)} u\left(\frac{y - b}{a} \right).$$

We can use Eq. (4.5.3) to produce the same result, as follows:

$$f_y(y) = \left[\{e^{-x} u(x)\} \left| \left| \frac{d}{dy} \frac{y - b}{a} \right| \right| \right] \Bigg|_{x=(y-b)/a} = \frac{1}{|a|} e^{-((y-b)/a)} u\left(\frac{y - b}{a} \right).$$

One check we have on this result is to use Eq. (4.4.3) and compare the result with $E\{Y\}$ using $f_y(y)$. From Eq. (4.4.3),

$$E\{aX + b\} = aE\{X\} + b = a + b$$

In many applications it is also necessary to find the joint pdf corresponding to transformations of jointly distributed random variables. For instance, consider the bivariate case where we are given two random variables X and Y with joint pdf $f_{XY}(x, y)$ and we wish to find the joint pdf of W and Z that are related to X and Y by $W = g(X, Y)$ and $Z = h(X, Y)$. Similar to the univariate case, we solve the given transformations for x and y in terms of z and w and use the formula

$$f_{WZ}(w, z) = \sum_i \frac{f_{XY}(x, y)}{|J(x, y)|} \Bigg|_{x=x_i, y=y_i}, \tag{4.5.4}$$

where x_i and y_i represent the simultaneous solutions of the given transformations and $J(x, y)$ is the Jacobian defined by the determinant

$$J(x, y) = \begin{vmatrix} \frac{\partial}{\partial x} g(x, y) & \frac{\partial}{\partial y} g(x, y) \\ \frac{\partial}{\partial x} h(x, y) & \frac{\partial}{\partial y} h(x, y) \end{vmatrix}. \tag{4.5.5}$$

The extension of this approach to more than two functions of more than two random variables is straightforward.

Example 4.5.2 Given the two random variables X and Y with joint pdf

$$f_{XY}(x, y) = \begin{cases} e^{-(x+y)}, & \text{for } x \geq 0, \quad y \geq 0 \\ 0, & \text{elsewhere,} \end{cases}$$

we wish to find the joint pdf of the random variables W and Z related to X and Y by the expressions

$$W = X + 2Y \quad \text{and} \quad Z = 2X + Y.$$

The Jacobian is

$$J = \begin{vmatrix} 1 & 2 \\ 2 & 1 \end{vmatrix} = [1 - 4] = -3$$

and solving for x and y yields

$$x = \frac{2z - w}{3} \quad \text{and} \quad y = \frac{2w - z}{3}.$$

Using Eq. (4.5.4), we obtain

$$
\begin{aligned}
f_{WZ}(w, z) &= \left. \frac{f_{XY}(x, y)}{|-3|} \right|_{x=(2z-w)/3, \, y=(2w-z)/3} \\
&= \frac{1}{3} \exp \left\{ -\left[\frac{2z - w}{3} + \frac{2w - z}{3} \right] \right\} \\
&= \frac{1}{3} e^{-[(z+w)/3]}, \quad w \geq 0, \quad \frac{w}{2} \leq z \leq 2w \\
&= 0, \quad \text{otherwise.}
\end{aligned}
$$

It is sometimes necessary to determine the pdf of one function of two random variables. In this case it is sometimes useful to define an auxiliary variable. For instance, given two random variables X and Y with joint pdf $f_{XY}(x, y)$ and the function $W = g(X, Y)$, suppose that we desire the pdf of W. In this situation, we can define the auxiliary variable Z by $Z = X$ or $Z = Y$ and use Eq. (4.5.4) to obtain the joint pdf of W and Z. We then integrate out the variable Z.

Example 4.5.3 Given the joint pdf of the random variables X and Y, we wish to find the pdf of $W = X + Y$. To facilitate the solution, we define $Z = Y$, so Eq. (4.5.5) yields.

$$J(x, y) = \begin{vmatrix} 1 & 1 \\ 0 & 1 \end{vmatrix} = 1.$$

Solving for x and y, we have $x = w - z$ and $y = z$, so Eq. (4.5.4) gives

$$f_{WZ}(w, z) = \frac{f_{XY}(x, y)}{|1|}\bigg|_{x=w-z, y=z} = f_{XY}(w - z, z).$$

Thus

$$f_W(w) = \int_{-\infty}^{\infty} f_{XY}(w - z, z)dz. \tag{4.5.6}$$

Note that a special case occurs when X and Y are independent. Then

$$f_W(w) = \int_{-\infty}^{\infty} f_X(w - z)f_Y(z)dz. \tag{4.5.7}$$

That is, the pdf of the sum of two independent random variables is the convolution of their marginal densities.

4.6   Special Distributions

There are several distribution functions that arise so often in communications problems that they deserve special attention. In this section we examine five such distributions: the uniform, Gaussian, and Rayleigh distributions for continuous random variables, and the binomial and Poisson distributions for discrete random variables.

The Uniform Distribution

A random variable X is said to have a uniform distribution if the pdf of X is given by

$$f_X(x) = \begin{cases} \frac{1}{b-a}, & a \leq x \leq b \\ 0, & \text{otherwise.} \end{cases} \tag{4.6.1}$$

The mean and variance of X are directly

$$E(X) = \int_a^b x \frac{1}{b - a} dx = \frac{b + a}{2} \tag{4.6.2}$$

and

$$\text{var}(X) = E(X^2) - E^2(X) = \frac{(b - a)^2}{12}. \tag{4.6.3}$$

The CDF of X is straightforwardly shown to be

$$F_X(x) = \int_a^x \frac{1}{b-a} d\lambda = \frac{1}{b-a} \int_a^x d\lambda$$

$$= \frac{x-a}{b-a}, \quad a \le x \le b$$

$$= 0, \quad x < a$$

$$= 1, \quad x > b. \tag{4.6.4}$$

The characteristic function of a uniform random variable is given by

$$\Phi_X(\omega) = \int_a^b \frac{1}{b-a} e^{j\omega x} dx = e^{j\omega[(a+b)/2]} \frac{\sin[(b-a)\omega/2]}{(b-a)\omega/2}. \tag{4.6.5}$$

The Gaussian Distribution

A random variable X is said to have a Gaussian or normal distribution if its pdf is of the form

$$f_X(x) = \frac{1}{\sqrt{2\pi}\sigma} e^{-(x-\mu)^2/2\sigma^2}, \quad -\infty < x < \infty. \tag{4.6.6}$$

We find that

$$E(X) = \frac{1}{\sqrt{2\pi}\sigma} \int_{-\infty}^{\infty} x e^{-(x-\mu)^2/2\sigma^2} dx.$$

Letting $y = (x-\mu)/\sigma$, we obtain

$$E(X) = \frac{1}{\sqrt{2\pi}} \int_{-\infty}^{\infty} (\sigma y + \mu) e^{-y^2/2} dy$$

$$= \frac{\sigma}{\sqrt{2\pi}} \int_{-\infty}^{\infty} y e^{-y^2/2} dy + \frac{\mu}{\sqrt{2\pi}} \int_{-\infty}^{\infty} e^{-y^2/2} dy$$

$$= 0 + \mu = \mu. \tag{4.6.7}$$

Furthermore,

$$E(X^2) = \frac{1}{\sqrt{2\pi}\sigma} \int_{-\infty}^{\infty} x^2 e^{-(x-\mu)^2/2\sigma^2} dx.$$

Again letting $y = (x-\mu)/\sigma$, we have that

$$E(X^2) = \frac{1}{\sqrt{2\pi}} \int_{-\infty}^{\infty} (\sigma y + \mu)^2 e^{-y^2/2} dy$$

$$= \frac{1}{\sqrt{2\pi}} \left\{ \sigma^2 \left[y e^{-y^2/2} \Big|_{-\infty}^{\infty} + \int_{-\infty}^{\infty} e^{-y^2/2} dy \right] \right.$$

$$+2\mu\sigma \int_{-\infty}^{\infty} y e^{-y^2/2} dy + \mu^2 \int_{-\infty}^{\infty} e^{-y^2/2} dy \Big\}$$
$$= \sigma^2 + \mu^2.$$
(4.6.8)

Therefore,

$$\text{var}(X) = \sigma^2 + \mu^2 - \mu^2 = \sigma^2.$$
(4.6.9)

The CDF of X is written as

$$F_X(x) = \frac{1}{\sqrt{2\pi}\sigma} \int_{-\infty}^{x} e^{-(y-\mu)^2/2\sigma^2} dy$$
$$= \frac{1}{\sqrt{2\pi}} \int_{-\infty}^{(x-\mu)/\sigma} e^{-z^2/2} dz = \frac{1}{2} + \text{erf}\left[\frac{x-\mu}{\sigma}\right],$$
(4.6.10)

where $\text{erf}\left[\frac{x-\mu}{\sigma}\right]$ is the error function, defined by

$$\text{erf}(x) = \frac{1}{\sqrt{2\pi}} \int_{0}^{x} e^{-\gamma^2/2} d\gamma.$$
(4.6.11)

The reader is cautioned that several different definitions of the error function exist, and hence the exact definition of $\text{erf}(x)$ should be checked before tables or other results are used.

The characteristic function of a Gaussian pdf is

$$\Phi_X(\omega) = \int_{-\infty}^{\infty} \frac{1}{\sqrt{2\pi}\sigma} e^{j\omega x} e^{-(x-\mu)^2/2\sigma^2} dx$$
$$= \frac{1}{\sqrt{2\pi}\sigma} \int_{-\infty}^{\infty} e^{-(x^2 - 2[j\omega\sigma^2 + \mu]x + \mu^2)/2\sigma^2} dx.$$

We can complete the square in the exponent by adding and subtracting $\left[-\omega^2\sigma^4 + 2j\omega\sigma^2\mu\right]/2\sigma^2$ to obtain

$$\Phi_X(\omega) = \left[e^{j\omega\mu - \omega^2\sigma^2/2}\right] \frac{1}{\sqrt{2\pi}\sigma} \int_{-\infty}^{\infty} e^{-[x+(j\omega\sigma^2 + \mu)]^2/2\sigma^2} dx$$
$$= e^{j\omega\mu - \omega^2\sigma^2/2},$$
(4.6.12)

since the integral is $\sqrt{2\pi}\sigma$. It is important to note that the Gaussian pdf is completely determined by specifying the mean and variance.

Multivariate Gaussian distributions also occur in many applications. We briefly consider here the bivariate Gaussian distribution. Two random variables X and Y are said to be jointly Gaussian distributed or jointly normal if their joint pdf is given by

$$f_{XY}(x, y) = \frac{1}{2\pi\sigma_X\sigma_Y\sqrt{1 - \rho_{XY}^2}} \exp\left\{\frac{-1}{2(1 - \rho_{XY}^2)}\left[\left(\frac{x - \mu_X}{\sigma_X}\right)^2\right.\right.$$

$$\left.\left. -2\rho_{XY}\frac{x - \mu_X}{\sigma_X}\frac{y - \mu_Y}{\sigma_Y} + \left(\frac{y - \mu_Y}{\sigma_Y}\right)^2\right]\right\}, \tag{4.6.13}$$

where ρ_{XY} is the correlation coefficient defined in Eq. (4.4.16). Note that if $\rho_{XY} = 0$ in Eq. (4.6.13), then $f_{XY}(x, y) = f_X(x)f_Y(y)$, and hence X and Y are independent. This result is an important property of Gaussian random variables; namely, that if two Gaussian random variables are uncorrelated, they are also independent. This property does not hold in general for non-Gaussian random variables. The characteristic function corresponding to a bivariate Gaussian distribution is

$$\Phi_{XY}(\omega_1, \omega_2) = \exp\left\{j[\omega_1\mu_X + \omega_2\mu_Y] - \frac{1}{2}[\sigma_X^2\omega_1^2 + 2\omega_1\omega_2\rho_{XY}\sigma_X\sigma_Y + \sigma_Y^2\omega_2^2]\right\}. \tag{4.6.14}$$

Note that if we are given n independent Gaussian random variables $X_i, i = 1, 2, \ldots, n$, and we form $Z = \sum_{i=1}^{n} X_i$ then

$$\Phi_Z(\omega) = \prod_{i=1}^{n} \Phi_{X_i}(\omega) = e^{j\omega\sum_{i=1}^{n}\mu_{X_i} - (\omega^2/2)\sum_{i=1}^{n}\sigma_{X_i}^2}$$

$$= e^{j\omega\mu_Z - \omega^2\sigma_Z^2/2}, \tag{4.6.15}$$

so that Z is also Gaussian. A form of this result can be extended to dependent Gaussian random variables to yield the conclusion that any linear combination of Gaussian random variables is also Gaussian.

The Rayleigh Distribution

A random variable X is said to have a Rayleigh distribution if its pdf is of the form

$$f_X(x) = \frac{x}{\alpha^2}e^{-x^2/2\alpha^2}u(x). \tag{4.6.16}$$

The mean and variance of the Rayleigh distribution are given by

$$E(X) = \alpha\sqrt{\frac{\pi}{2}} \tag{4.6.17}$$

and

$$\text{var}(X) = \left(2 - \frac{\pi}{2}\right)\alpha^2, \tag{4.6.18}$$

respectively.

The Binomial Distribution

The binomial distribution arises when we are considering n independent trials of an experiment admitting only two possible outcomes with constant probabilities p and $1 - p$, respectively. Thus, if the probability of "success" on any one trial is p, the probability of exactly x successes is

$$P[X = x] = b(x; n, p) = \binom{n}{x} p^x (1 - p)^{n-x} \tag{4.6.19}$$

for $x = 0, 1, 2, \ldots, n$. The notation $b(x; n, p)$ is fairly standard in the literature. From the binomial expansion we see that this density sums to 1, since

$$\sum_{x=0}^{n} \binom{n}{x} p^x (1 - p)^{n-x} = (p + 1 - p)^n = 1. \tag{4.6.20}$$

The expected value of X follows as

$$E(X) = \sum_{x=0}^{n} x \binom{n}{x} p^x (1 - p)^{n-x}$$

$$= \sum_{x=1}^{n} np \binom{n-1}{x-1} p^{x-1} (1 - p)^{n-1-(x-1)}$$

$$= np \sum_{y=0}^{n-1} \binom{n-1}{y} p^y (1 - p)^{n-1-y}$$

$$= np(p + 1 - p)^{n-1} = np. \tag{4.6.21}$$

To determine the variance, we find $E\{X(X-1)\}$ using manipulations similar to those in Eq. (4.6.21), that is,

$$E\{X(X-1)\} = \sum_{x=0}^{n} x(x-1) \binom{n}{x} p^x (1 - p)^{n-x}$$

$$= 0 + 0 + \sum_{x=2}^{n} n(n-1) p^2 \binom{n-2}{x-2} p^{x-2} (1 - p)^{n-2-(x-2)}$$

$$= n(n-1) p^2 \sum_{y=0}^{n-2} \binom{n-2}{y} p^y (1 - p)^{n-2-y}$$

$$= n(n-1) p^2 = E(X^2) - E(X). \tag{4.6.22}$$

Now since $E(X) = np$,

$$\text{var}(X) = E(X^2) - E^2(X) = n(n-1) p^2 + np - n^2 p^2 = np(1 - p). \tag{4.6.23}$$

The characteristic function of X can be evaluated straightforwardly as

$$\Phi_X(\omega) = E\left\{e^{j\omega X}\right\} = \sum_{x=0}^{n} e^{j\omega x} \binom{n}{x} p^x (1-p)^{n-x}$$

$$= \sum_{x=0}^{n} \binom{n}{x} \left(pe^{j\omega}\right)^x (1-p)^{n-x} = \left(pe^{j\omega} + 1 - p\right)^n. \qquad (4.6.24)$$

The moments of a binomial random variable X are usually evaluated from its characteristic function.

Note that unlike a continuous random variable, the equality sign in the CDF is of the utmost importance for discrete random variables. For example, for a binomial random variable X,

$$F_X(\alpha) = P[X \le \alpha] = \sum_{x=0}^{\alpha} b(x; n, p) = \sum_{x=0}^{\alpha} \binom{n}{x} p^x (1-p)^{n-x} \qquad (4.6.25)$$

while

$$P[X < \alpha] = \sum_{x=0}^{\alpha-1} \binom{n}{x} p^x (1-p)^{n-x}. \qquad (4.6.26)$$

The $x = \alpha$ term is missing in Eq. (4.6.26). There are extensive tables of the binomial pdf and CDF.

The Poisson Distribution

The Poisson distribution can be derived as a limiting case of the binomial distribution when n is large and p is small, and Poisson random variables play an important role in queueing problems. A random variable X is said to have a Poisson distribution if its pdf is given by

$$P[X = x] = f(x; \lambda) = \frac{\lambda^x}{x!} e^{-\lambda}, \qquad x = 0, 1, \ldots. \qquad (4.6.27)$$

It is easy to show that $f(x; \lambda)$ sums to 1 as

$$\sum_{x=0}^{\infty} f(x; \lambda) = \sum_{x=0}^{\infty} \frac{\lambda^x}{x!} e^{-\lambda} = e^{-\lambda} \sum_{x=0}^{\infty} \frac{\lambda^x}{x!} = e^{-\lambda}\left(e^{\lambda}\right) = 1. \qquad (4.6.28)$$

The mean of X is

$$E(X) = \sum_{x=0}^{\infty} x \frac{\lambda^x}{x!} e^{-\lambda} = 0 + \sum_{x=1}^{\infty} \lambda e^{-\lambda} \frac{\lambda^{x-1}}{(x-1)!}$$

$$= \lambda e^{-\lambda} \sum_{y=0}^{\infty} \frac{\lambda^y}{y!} = \lambda. \qquad (4.6.29)$$

Similarly, we can show that

$$E\{X(X-1)\} = \lambda^2,$$

So

$$\text{var}(X) = E\{X(X-1)\} + E(X) - E^2(X) = \lambda^2 + \lambda - \lambda^2 = \lambda. \tag{4.6.30}$$

The characteristic function of a Poisson random variable is

$$\Phi_X(\omega) = \sum_{x=0}^{\infty} e^{j\omega x} \frac{\lambda^x}{x!} e^{-\lambda} = \sum_{x=0}^{\infty} \frac{\left(e^{j\omega}\lambda\right)^x}{x!} e^{-\lambda}$$
$$= e^{-\lambda} e^{\lambda e^{j\omega}} = e^{\lambda\left(e^{j\omega}-1\right)}. \tag{4.6.31}$$

4.7   Stochastic Processes and Correlation

Random variables are defined on the outcomes of a random experiment. If we perform the random experiment, we obtain a value or range of values of the random variable. For random processes or stochastic processes, however, the situation is quite different. In the case of a random process, the outcome of the random experiment selects a particular function of time, called a realization of the stochastic process. Thus a random or stochastic process depends both on the outcome of a random experiment and on time.

Definition 4.7.1 A random process or stochastic process $X(\xi, t)$ is a family of random variables indexed by the parameter $t \in T$, where T is called the index set.

A simple example of a random process is a single frequency sinusoid with random amplitude $A(\xi)$, so that

$$X(\xi, t) = A(\xi) \cos \omega_c t \tag{4.7.1}$$

for $-\infty < t < \infty$. A few realizations of this random process corresponding to outcomes of the random experiment are sketched in Fig. 4.5. Note that once we specify $\xi = \xi_1$ (say), we have a deterministic time function $X(\xi_1, t) = A(\xi_1) \cos \omega_c t$, since $A(\xi_1)$ is just a number. Further, if we evaluate $X(\xi, t)$ at some specific value of t, say $t = t_1$, we are left with a random variable

$$X(\xi, t_1) = A(\xi) \cos \omega_c t_1, \tag{4.7.2}$$

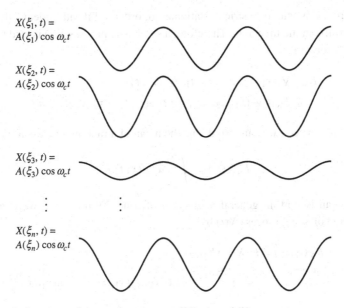

$$X(\xi_1, t) = A(\xi_1) \cos \omega_c t$$

$$X(\xi_2, t) = A(\xi_2) \cos \omega_c t$$

$$X(\xi_3, t) = A(\xi_3) \cos \omega_c t$$

$$\vdots$$

$$X(\xi_n, t) = A(\xi_n) \cos \omega_c t$$

Fig. 4.5 Sample functions of the random process $X(\xi, t) = A(\xi) \cos \omega_c t$

which is no longer a function of time. Although we have only argued these points for the particular random process in Eq. (4.7.1), the conclusions are true for more general random processes.

Since a stochastic process is a function of t, the probability distributions and densities of the stochastic process also depend on t. For example, the first-order pdf of a random process $X(\xi, t)$ may be written as $f_X(x; t)$, where the explicit dependence on ξ is dropped. Often, the dependence on ξ is not expressed in $X(\xi, t)$, so that it is common to write $X(\xi, t) = X(t)$ for a stochastic process. We can write an expression for the CDF of $X(t)$ as

$$F_X(x; t) = P[X(t) \leq x] = \int_{-\infty}^{x} f_X(\alpha; t)d\alpha. \tag{4.7.3}$$

Joint distributions of the random variables obtained by evaluating $X(t)$ at $t_1, t_2, ...,$ and t_n, are also quite important in characterizing the random process. The nth-order distribution of a random process $X(t)$ is given by

$$F_{X_1...X_n}(x_1, x_2, ..., x_n; t_1, t_2, ..., t_n)$$
$$= P[X(t_1) \leq x_1, X(t_2) \leq x_2, ..., X(t_n) \leq x_n]$$
$$= \int_{-\infty}^{x_n} \cdots \int_{-\infty}^{x_2} \int_{-\infty}^{x_1} f_{X_1 \cdots X_n}(\alpha_1, \alpha_2, ..., \alpha_n; t_1, t_2, ..., t_n)d\alpha_1 \, d\alpha_2 \cdots d\alpha_n.$$

$$\tag{4.7.4}$$

A random process is said to be strictly stationary of order n if its nth-order CDF and pdf are invariant to shifts in the time axis. Therefore, a stochastic process $X(t)$ is said to be strictly stationary of order n if

$$F_{X_1 \cdots X_n}(x_1, x_2, \ldots, x_n; t_1, t_2, \ldots, t_n)$$
$$= F_{X_1 \cdots X_n}(x_1, x_2, \ldots, x_n; t_1 + \tau, t_2 + \tau, \ldots, t_n + \tau). \qquad (4.7.5)$$

As one might expect, we can also define the mean of a random process $X(t)$ by

$$E\{X(t)\} = \int_{-\infty}^{\infty} x f_X(x; t) dx, \qquad (4.7.6)$$

where the mean is now in general a function of time. Furthermore, we can define the autocorrelation of a real process $X(t)$ by

$$R_X(t_1, t_2) = E\{X(t_1)X(t_2)\}$$
$$= \int_{-\infty}^{\infty} \int_{-\infty}^{\infty} x_1 x_2 f_{X_1 X_2}(x_1, x_2; t_1, t_2) dx_1 dx_2 \qquad (4.7.7)$$

and the covariance as

$$\text{cov}[X(t_1), X(t_2)] = R_X(t_1, t_2) - E[X(t_1)]E[X(t_2)]. \qquad (4.7.8)$$

As with random variables, these quantities are very good "summary indicators" of random process behavior.

Another important form of stationarity does not require that we examine the nth-order CDF or pdf of the process, but we compute only its mean and autocorrelation. A stochastic process is said to be wide-sense stationary (WSS) or weakly stationary if

$$E\{X(t_1)\} = E\{X(t_2)\} = \text{constant}, \qquad (4.7.9)$$

and

$$R_X(t_1, t_2) = E\{X(t_1)X(t_2)\} = E\{X(t)X(t + |t_2 - t_1|)\}$$
$$= E\{X(t)X(t + \tau)\} = R_X(\tau). \qquad (4.7.10)$$

Thus a random process is WSS if its mean is a constant and its autocorrelation depends only on the absolute time difference $|t_2 - t_1|$.

Example 4.7.1 Consider the random process.

$$X(t) = A \cos \omega_c t, \quad -\infty < t < \infty, \qquad (4.7.11)$$

where ω_c is a constant and A is a zero-mean Gaussian random variable with variance σ_A^2. We would like to find the mean, autocorrelation, and the first-order pdf of $X(t)$.

For the mean we find that

$$E\{X(t)\} = E\{A\}\cos \omega_c t = 0, \tag{4.7.12}$$

and the autocorrelation is

$$E\{X(t_1)X(t_2)\} = E\{A^2\}\cos \omega_c t_1 \cos \omega_c t_2$$

$$= \frac{\sigma_A^2}{2}[\cos \omega_c(t_1 + t_2) + \cos \omega_c(t_1 - t_2)]. \tag{4.7.13}$$

Letting $t_1 = t_2 = t$, we find that

$$E\{X^2(t)\} = \sigma_A^2\left[\frac{1}{2} + \frac{1}{2}\cos 2\omega_c t\right] = \text{var}\{X(t)\}, \tag{4.7.14}$$

since $X(t)$ is zero mean from Eq. (4.7.12). Therefore, the first-order pdf of $X(t)$ is Gaussian with zero mean and variance given by Eq. (4.7.14).

We can also ask if this process is wide-sense stationary. Certainly, the mean of $X(t)$ is a constant (0), but from Eq. (4.7.13), it is clear that $R_X(t_1, t_2)$ is not a function of $|t_2 - t_1|$ only. Hence the random process is not WSS.

There are several properties of a WSS stochastic process $X(t)$ that are useful in applications. In stating these properties, we use the notation in Eq. (4.7.10) that $R_X(\tau) = E\{X(t)X(t + \tau)\}$.

Property 1 The mean-squared value of $X(t)$ is

$$R_X(0) = E\{X(t)X(t + \tau)\}|_{\tau=0} = E\{X^2(t)\}. \tag{4.7.15}$$

Property 2 For a WSS process, the autocorrelation is an even function of τ, since

$$R_X(\tau) = E\{X(t)X(t + \tau)\} = E\{X(\alpha - \tau)X(\alpha)\} = R_X(-\tau), \tag{4.7.16}$$

where we let $t = \alpha - \tau$.

Property 3 For a WSS process, the autocorrelation is a maximum at the origin. This can be shown by considering the quantity

$$E\{[X(t + \tau) \pm X(t)]^2\} = E\{X^2(t + \tau)\} + E\{X^2(t)\} \pm 2E\{X(t)X(t + \tau)\}$$

$$= 2R_X(0) \pm 2R_X(\tau). \tag{4.7.17}$$

However, we know that $E\{[X(t + \tau) \pm X(t)]^2\} \geq 0$, so

$$R_X(0) \geq |R_X(\tau)|. \tag{4.7.18}$$

Given two real random processes $X(t)$ and $Y(t)$, we can define two cross-correlation functions as

$$R_{XY}(t_1, t_2) = E\{X(t_1)Y(t_2)\} \tag{4.7.19}$$

and

$$R_{YX}(t_1, t_2) = E\{Y(t_1)X(t_2)\}. \tag{4.7.20}$$

The cross-correlation function does not have the same properties as the auto–correlation, but the following properties of the cross-correlation for WSS processes can prove useful.

Property 4 For two WSS processes with

$$R_{XY}(\tau) = E\{X(t)Y(t+\tau)\} \tag{4.7.21}$$

and

$$R_{YX}(\tau) = E\{Y(t)X(t+\tau)\}, \tag{4.7.22}$$

we have that

$$R_{XY}(-\tau) = R_{YX}(+\tau). \tag{4.7.23}$$

To show this, we begin with Eq. (4.7.21) and consider $R_{XY}(-\tau)$,

$$R_{XY}(-\tau) = E\{X(t)Y(t-\tau)\} = E\{X(\alpha+\tau)Y(\alpha)\} = R_{YX}(\tau), \tag{4.7.24}$$

where we have defined $\alpha = t - \tau$.

Property 5 For two WSS processes $X(t)$ and $Y(t)$, the cross-correlation is bounded as

$$|R_{XY}(\tau)| \leq \frac{1}{2}[R_X(0) + R_Y(0)]. \tag{4.7.25}$$

To demonstrate this, we expand the nonnegative quantity

$$E\{[X(t) \pm Y(t+\tau)]^2\} \geq 0 \tag{4.7.26}$$

to obtain

$$R_X(0) \pm 2R_{XY}(\tau) + R_Y(0) \geq 0, \tag{4.7.27}$$

thus directly yielding Eq. (4.7.25).

Property 6 For two WSS random processes $X(t)$ and $Y(t)$, we can write

$$E\left\{\left[\frac{X(t+\tau)}{\sqrt{R_X(0)}} - \frac{Y(t)}{\sqrt{R_Y(0)}}\right]^2\right\} = \frac{E\{X^2(t+\tau)\}}{R_X(0)} + \frac{E\{Y^2(t)\}}{R_Y(0)} - \frac{2E\{X(t+\tau)Y(t)\}}{\sqrt{R_X(0)R_Y(0)}} \geq 0.$$

(4.7.28)

Now, the first two terms in Eq. (4.7.28) each have the value 1, so

$$\frac{2R_{XY}(\tau)}{\sqrt{R_X(0)R_Y(0)}} \leq 2$$

or

$$R_{XY}^2(\tau) \leq R_X(0)R_Y(0)$$

(4.7.29)

Property 7 If $Z(t) = X(t) + Y(t)$, where $X(t)$ and $Y(t)$ are WSS, then

$$R_Z(\tau) = R_X(\tau) + R_Y(\tau) + R_{XY}(\tau) + R_{YX}(\tau).$$

(4.7.30)

The proof is direct.

Property 8 If a WSS random process $X(t)$ is periodic of period T, its auto–correlation is also periodic with period T

To demonstrate this, let $X(t) = X(t+T) = X(t+nT)$ for integer n. Then

$$R_X(\tau) = E\{X(t)X(t+\tau)\} = E\{X(t)X(t+\tau+nT)\}$$
$$= R_X(\tau+nT).$$

(4.7.31)

4.8 Ergodicity

The concept of stationarity, particularly in the wide sense, is extremely important to the analysis of random processes that occur in communication systems. However, for a specific application, it often occurs that the (ensemble) mean or autocorrelation function is unknown and must be estimated. The most common way of estimating these quantities in engineering applications is to observe a particular sample function of the random process and then compute the mean, autocorrelation, or other required average for this specific sample function. In such a situation, we are actually computing time averages over the (single) sample function.

For example, given a random process $X(\xi, t)$, if we observe a particular sample function of this process, we thus have $X(\xi_1, t) = X(t)$, which is now only a function of t. For this sample function, we can compute its average value as

$$\langle X(\xi_1, t) \rangle = \langle x(t) \rangle = \lim_{T \to \infty} \frac{1}{2T} \int_{-T}^{T} x(t) \, dt, \qquad (4.8.1)$$

where $\langle X(\xi_1, t) \rangle$ denotes averaging over time. We can also define a (time) autocorrelation function for this sample function as

$$\mathcal{R}_X(\tau) = \langle x(t)x(t + \tau) \rangle = \lim_{T \to \infty} \frac{1}{2T} \int_{-T}^{T} x(t)x(t + \tau) \, dt. \qquad (4.8.2)$$

Now, if we use these time averages as estimates of the ensemble mean and autocorrelation, which we discussed previously, we are claiming that

$$\langle x(t) \rangle = \lim_{T \to \infty} \frac{1}{2T} \int_{-T}^{T} x(t) \, dt = E\{X(\xi; t)\}$$

$$= \int_{-\infty}^{\infty} x(\xi; t) f_X(x, \xi; t) \, dx(\xi; t) \qquad (4.8.3)$$

and

$$\mathcal{R}_X(\tau) = \langle x(t)x(t + \tau) \rangle = E\{X(\xi; t)X(\xi; t + \tau)\}$$

$$= \int_{-\infty}^{\infty} \int_{-\infty}^{\infty} x_1(\xi; t)x_2(\xi; t + \tau) f_{X_1 X_2}(x_1, x_2; t, t + \tau) \, dx_1(\xi; t) \, dx_2(\xi; t + \tau). \qquad (4.8.4)$$

Equation (4.8.3) asserts that the time average of $x(t)$ equals the ensemble average of $X(\xi; t)$. Equation (4.8.4) makes a similar assertion for time and ensemble autocorrelations. Now, time averages and ensemble averages are fundamentally quite different quantities, and when they are the same, we have a very special situation. In fact, we give processes with this property a special name.

Definition 4.8.1 We call random processes with the property that time averages and ensemble averages are equal *ergodic*.

Ergodicity does not just apply to the mean and autocorrelation—it applies to all time and ensemble averages. Thus, if we have an ergodic random process, if we can determine the ensemble averages by computing the corresponding time average over a particular realization of the random process. Clearly, this is a powerful tool, and ergodicity is often assumed in engineering applications. For a much more careful discussion and development of the ergodic property of random processes, see [1].

We note that for a random process to be ergodic, it must first be stationary. This result is intuitive, since if a time average is to equal an ensemble average, the ensemble average cannot be changing with time.

4.9   Spectral Densities

Electrical engineers are accustomed to thinking about systems in terms of their frequency response and about signals in terms of their frequency content. We would like to be able to extend this "frequency-domain" thinking to incorporate stochastic processes. Unfortunately, the standard Fourier integral of a sample function of a stochastic process may not exist, and hence we must look elsewhere for a frequency-domain representation of random processes. For the restricted class of wide-sense-stationary random processes, it is possible to define a quantity called the power spectral density, or spectral density function, as the Fourier transform of the autocorrelation, so

$$S_X(\omega) = \mathcal{F}\{R_X(\tau)\} = \int_{-\infty}^{\infty} R_X(\tau)e^{-j\omega\tau}d\tau. \tag{4.9.1}$$

The power spectral density is an indicator of the distribution of signal power as a function of frequency. Since the Fourier transform is unique, we can also obtain the autocorrelation function of the WSS process $\{X(t), -\infty < t < \infty\}$ as

$$R_X(\tau) = \mathcal{F}^{-1}\{S_X(\omega)\} = \frac{1}{2\pi} \int_{-\infty}^{\infty} S_X(\omega)e^{j\omega\tau}d\omega. \tag{4.9.2}$$

Equations (4.9.1) and (4.9.2) are sometimes called the Wiener-Khintchine relations. We now consider two simple examples.

Example 4.9.1 We are given a WSS stochastic process $\{X(t), -\infty < t < \infty\}$ with zero mean and $R_X(\tau) = A \cos \omega_c\tau$ for $-\infty < \tau < \infty$. We desire the power spectral density of $X(t)$. Directly from Eq. (4.9.1) we find that.

$$\begin{aligned} S_X(\omega) &= \mathcal{F}\{R_X(\tau)\} = \mathcal{F}\{A \cos \omega_c\tau\} \\ &= \pi A[\delta(\omega + \omega_c) + \delta(\omega - \omega_c)]. \end{aligned} \tag{4.9.3}$$

Thus a WSS stochastic process with a purely sinusoidal autocorrelation function with period $T_c = 2\pi / \omega_c$ has a spectral density function with impulses at $\pm\omega_c$.

Example 4.9.2 A WSS stochastic process $Y(t)$ has the spectral density function

$$S_Y(\omega) = \tau_0 \frac{\sin^2(\omega\tau_0/2)}{(\omega\tau_0/2)^2}. \tag{4.9.4}$$

We wish to find the autocorrelation function of the process. Since by Eq. (4.9.2), $R_Y(\tau) = \mathcal{F}^{-1}\{S_Y(\omega)\}$, we have from tables that

$$R_Y(\tau) = \begin{cases} 1 - \frac{|\tau|}{\tau_0}, & |\tau| < \tau_0 \\ 0, & |\tau| > \tau_0. \end{cases} \tag{4.9.5}$$

There are several simple properties of the power spectral density that we state here.

Property 1 The mean-square value of a WSS process is given by

$$E\{X^2(t)\} = R_X(0) = \frac{1}{2\pi}\int_{-\infty}^{\infty} S_X(\omega)\,d\omega. \tag{4.9.6}$$

Property 2 The spectral density function of a WSS process is always nonnegative,

$$S_X(\omega) \geq 0 \quad \text{for all } \omega. \tag{4.9.7}$$

Property 3 The power spectral density of a real WSS random process is an even function of ω:

$$S_X(\omega) = S_X(-\omega). \tag{4.9.8}$$

Property 4 The value of the spectral density function at $\omega = 0$ is [from Eq. (4.9.1)].

$$S_X(0) = \int_{-\infty}^{\infty} R_X(\tau)\,d\tau. \tag{4.9.9}$$

Of course, in communication systems it is often necessary to determine the response of a given linear filter to a random process input. It is problems of this type for which the utility of the power spectral density is most apparent. Specifically, given a linear system with impulse response $h(t)$ and transfer function $H(\omega)$, suppose that we wish to find the power spectral density at its output when its input is a WSS random process $X(t)$ with zero mean and autocorrelation function $R_x(\tau)$. The output stochastic process can be expressed as $Y(t) = \int_{-\infty}^{\infty} h(\alpha)X(t-\alpha)\,d\alpha$, so

$$R_Y(\tau) = E\{Y(t)Y(t+\tau)\}$$

$$= \int_{-\infty}^{\infty} \int_{-\infty}^{\infty} h(\alpha) h(\beta) E\{X(t-\alpha) X(t+\tau-\beta)\} \, d\alpha \, d\beta$$

$$= \int_{-\infty}^{\infty} \int_{-\infty}^{\infty} h(\alpha) h(\beta) R_X(\tau+\alpha-\beta) \, d\alpha \, d\beta. \tag{4.9.10}$$

Working first with the integral on β, we note that

$$\int_{-\infty}^{\infty} h(\beta) R_X(\tau+\alpha-\beta) \, d\beta = h(\tau+\alpha) * R_X(\tau+\alpha) \triangleq g(\tau+\alpha). \tag{4.9.11}$$

Using Eq. (4.9.11), we rewrite Eq. (4.9.10) as

$$R_Y(\tau) = \int_{-\infty}^{\infty} h(\alpha) g(\tau+\alpha) \, d\alpha$$

$$= \int_{-\infty}^{\infty} h(-\lambda) g(\tau-\lambda) \, d\lambda = h(-\tau) * g(\tau), \tag{4.9.12}$$

where we have made the change of variables that $\alpha = -\lambda$. Finally, substituting for $g(\tau)$, we obtain

$$R_Y(\tau) = h(-\tau) * h(\tau) * R_X(\tau). \tag{4.9.13}$$

Taking the Fourier transform of both sides of Eq. (4.9.13), we get the useful and simple result

$$S_Y(\omega) = H^*(\omega) H(\omega) S_X(\omega) = |H(\omega)|^2 S_X(\omega). \tag{4.9.14}$$

Thus, given the system transfer function and the power spectral density of the input process, it is straightforward to obtain the power spectral density of the output process via Eq. (4.9.14).

Example 4.9.3 A WSS stochastic process $X(t)$ has zero mean and autocorrelation function $R_X(\tau) = V\delta(\tau)$. If $X(t)$ is applied to the input of a linear filter with transfer function $H(\omega) = 1/(1+j\omega RC)$, we wish to find the power spectral density of the output process $Y(t)$. Directly, we have (a process with a flat power spectral density is said to be white)

$$S_X(\omega) = V \quad \text{for} \quad -\infty < \omega < \infty$$

and

$$|H(\omega)|^2 = \frac{1}{1+\omega^2 R^2 C^2}, \tag{4.9.15}$$

so

$$S_Y(\omega) = \frac{V}{1 + \omega^2 R^2 C^2}, \quad -\infty < \omega < \infty. \tag{4.9.16}$$

We can also find the autocorrelation function of the output by taking the inverse transform of Eq. (4.9.16) to yield

$$R_Y(\tau) = \frac{V}{2RC} e^{-|\tau|/RC} \tag{4.9.17}$$

for $-\infty < \tau < \infty$.

We can define cross spectral densities of two WSS random processes $X(t)$ and $Y(t)$ as the Fourier transform of their cross-correlation function,

$$S_{XY}(\omega) = \mathcal{F}\{R_{XY}(\tau)\} = \int_{-\infty}^{\infty} R_{XY}(\tau) e^{-j\omega\tau} d\tau. \tag{4.9.18}$$

course, given $S_{xy}(\omega)$, we have

$$R_{XY}(\tau) = \mathcal{F}^{-1}\{S_{XY}(\omega)\} = \frac{1}{2\pi} \int_{-\infty}^{\infty} S_{XY}(\omega) e^{j\omega\tau} d\omega. \tag{4.8.19}$$

Given a linear system with impulse response $h(t)$ and transfer function $H(\omega)$, we find that it is of considerable practical interest to calculate the cross spectral density between the input process $X(t)$ and the output process $Y(t)$. Forming the cross correlation between the WSS process $X(t)$ and $Y(t)$, we have

$$R_{XY}(\tau) = E\{X(t)Y(t+\tau)\} = \int_{-\infty}^{\infty} h(\alpha) E\{X(t)X(t+\tau-\alpha)\}\, d\alpha$$

$$= \int_{-\infty}^{\infty} h(\alpha) R_X(\tau-\alpha)\, d\alpha = h(\tau) * R_X(\tau). \tag{4.9.20}$$

Using Fourier transform properties, we find that the desired cross spectral density is

$$S_{XY}(\omega) = H(\omega) S_X(\omega). \tag{4.9.21}$$

Equation (4.9.21) and its many variants are often valuable when it is necessary to identify an unknown system transfer function.

The Wiener-Khintchine relations in Eqs. (4.9.1) and (4.9.2) can also be developed by starting with a finite sample of a real random process, say $\{X(t), 0 < t < T\}$. We start by defining the sample spectral density function

$$S_{X_T}(\omega) = \frac{1}{T} \left| \int_0^T X(t) e^{-j\omega t} dt \right|^2 \tag{4.9.22}$$

and the sample autocorrelation function

$$R_{X_T}(\tau) = \begin{cases} \frac{1}{2\pi T} \int_0^{T-\tau} X(t)X(t+\tau)\, dt, & 0 \le \tau < T \\ 0, & \tau > T \\ R_{X_T}(-\tau), & \tau < 0, \end{cases} \tag{4.9.23}$$

which can be shown to be a Fourier transform pair,

$$S_{X_T}(\omega) = \int_{-T}^{T} R_{X_T}(\tau)e^{-j\omega\tau}\, d\tau \tag{4.9.24}$$

and

$$R_{X_T}(\tau) = \frac{1}{2\pi} \int_{-\infty}^{\infty} S_{X_T}(\omega)e^{j\omega\tau}\, d\omega \tag{4.9.25}$$

for $-\infty < \tau < \infty$. Taking the expectation of both sides of Eq. (4.9.25) yields

$$E\{R_{X_T}(\tau)\} = \frac{1}{2\pi} \int_{-\infty}^{\infty} E\{S_{X_T}(\omega)\}e^{j\omega\tau}\, d\omega, \tag{4.9.26}$$

$-\infty < \tau < \infty$. Now, for a zero mean, WSS stochastic process, it can be shown that

$$\lim_{T \to \infty} E\{R_{X_T}(\tau)\} = R_X(\tau), \tag{4.9.27}$$

where $R_x(t)$ is given by Eq. (4.9.2), assuming that the spectral density exists. Furthermore, it can also be demonstrated that

$$S_X(\omega) = \lim_{T \to \infty} \frac{1}{T} E\left\{ \left| \int_0^T X(t)e^{-j\omega t}\, dt \right|^2 \right\}. \tag{4.9.28}$$

Equations (4.9.27) and (4.9.28) can sometimes be used to advantage in practical problems.

4.10 Cyclostationary Processes

A common model of transmitted sequences in digital communications systems is given by

$$X(t) = \sum_{n=-\infty}^{\infty} a_n p(t - nT_s), \tag{4.10.1}$$

where $p(t)$ is the pulse shape, T_s is the symbol duration, and $\{a_n\}$ is a WSS sequence with $E\{a_n\} = \mu_a$ and $E\{a_n a_m\} = E\{a_l a_{l+k}\} = R_a(k)$, $k = |n - m|$. We would like to find the power spectral density of $X(t)$. The mean of $X(t)$ is immediately available as

$$E[X(t)] = \mu_a \sum_{n=-\infty}^{\infty} p(t - nT_s), \tag{4.10.2}$$

and the autocorrelation is given by

$$R_X(t_1, t_2) = E[X(t_1)X(t_2)]$$

$$= \sum_{n=-\infty}^{\infty} \sum_{m=-\infty}^{\infty} E[a_n a_m] p(t_1 - nT_s) p(t_2 - mT_s)$$

$$= \sum_{k=-\infty}^{\infty} R_a(k) \sum_{n=-\infty}^{\infty} p(t_1 - nT_s) p(t_2 - (k + n)T_s). \tag{4.10.3}$$

From Eqs. (4.10.2) and (4.10.3) it is clear that the sequence $X(t)$ is not WSS. As a result, the power spectral density cannot be defined using Eq. (4.9.1).

Random processes that satisfy the relations

$$E[Y(t_1 + T)] = E[Y(t_1)] \tag{4.10.4}$$

and

$$R_Y(t_1 + T, t_2 + T) = R_Y(t_1, t_2) \tag{4.10.5}$$

are called *cyclostationary* because they are periodic in their time arguments [2]. We see from Eqs. (4.10.2) and (4.10.3) that the sequence $X(t)$ is a cyclostationary process.

Fortunately, $X(t)$ can be modified to obtain a WSS process by allowing a random time delay. Consider a new sequence

$$X(t) = \sum_{n=-\infty}^{\infty} a_n p(t - nT_s - \lambda), \tag{4.10.6}$$

where λ is a uniformly distributed random variable over $0 \leq t < T_s$ independent of a_n. Then

$$E[X(t)] = \sum_{n=-\infty}^{\infty} \mu_a E[p(t - nT_s - \lambda)]$$

$$= \mu_a \sum_{n=-\infty}^{\infty} \frac{1}{T_s} \int_0^{T_s} p(t - nT_s - \lambda) \, d\lambda$$

$$= \frac{\mu_a}{T_s} \sum_{n=-\infty}^{\infty} \int_{t-(n+1)T_s}^{t-nT_s} p(\alpha) \, d\alpha = \frac{\mu_a}{T_s} \int_{-\infty}^{\infty} p(t) \, dt, \tag{4.10.7}$$

which is a constant. Further,

$$R_X(t_1, t_2) = E[X(t_1)X(t_2)]$$

$$= \sum_{n=-\infty}^{\infty} \sum_{m=-\infty}^{\infty} E[a_n a_m] \int_0^{T_s} \frac{1}{T_s} p(t_1 - nT_s - \lambda) p(t_2 - mT_s - \lambda) \, d\lambda$$

$$= \sum_{k=-\infty}^{\infty} R_a(k) \frac{1}{T_s} \sum_{n=-\infty}^{\infty} \int_0^{T_s} p(t_1 - nT_s - \lambda) p(t_2 - (n+k)T_s - \lambda) \, d\lambda$$

$$= \frac{1}{T_s} \sum_{k=-\infty}^{\infty} R_a(k) \sum_{n=-\infty}^{\infty} \int_{t_1-(n+1)T_s}^{t_1-nT_s} p(\alpha) p(\alpha + \tau - kT_s) \, d\alpha$$

$$= \frac{1}{T_s} \sum_{k=-\infty}^{\infty} R_a(k) \int_{-\infty}^{\infty} p(t) p(t + \tau - kT_s) \, dt$$

$$= \frac{1}{T_s} \sum_{k=-\infty}^{\infty} R_a(k) \mathcal{R}_p(\tau - kT_s), \tag{4.10.8}$$

where $\tau = |t_2 - t_1|$ and

$$\mathcal{R}_p(\tau) = \int_{-\infty}^{\infty} p(t) p(t + \tau) \, dt. \tag{4.10.9}$$

Since $R_X(t_1, t_2) = R_X(|t_2 - t_1|)$ and $E[X(t)] = $ constant, $X(t)$ in Eq. (4.10.6) is WSS.

To simplify Eq. (4.10.8) further, assume that the a_n sequence is statistically independent (but not zero mean),

$$R_a(k) = E[a_n a_{n+k}] = \begin{cases} \mu_a^2, & k \neq 0 \\ \sigma_a^2 + \mu_a^2, & k = 0, \end{cases} \tag{4.10.10}$$

where $\sigma_a^2 = E[a_n^2] - \mu_a^2$. Then, Eq. (4.10.8) yields

$$R_X(\tau) = \frac{\sigma_a^2}{T_s} \mathcal{R}_p(\tau) + \frac{\mu_a^2}{T_s} \sum_{k=-\infty}^{\infty} \mathcal{R}_p(\tau - kT_s). \tag{4.10.11}$$

Using Eq. (4.10.9),

$$S_p(\omega) = \mathcal{F}\{\mathcal{R}_p(\tau)\} = |P(\omega)|^2, \tag{4.10.12}$$

where $P(\omega) = \mathcal{F}\{p(t)\}$, we can write several different useful expressions for the power spectral density. Taking the Fourier transform of Eq. (4.10.8), we get the general relationship

$$S_X(\omega) = \frac{1}{T_s} |P(\omega)|^2 R_a(0) + \frac{1}{T_s} \sum_{k=-\infty}^{\infty} R_a(k) |P(\omega)|^2 e^{-j\omega kT_s}$$

$$= \frac{|P(\omega)|^2}{T_s} \left\{ R_a(0) + 2 \sum_{k=1}^{\infty} R_a(k) \cos k\omega T_s \right\}. \tag{4.10.13}$$

Based on the assumptions in Eq. (4.10.10), we can start with Eq. (4.10.11) rewritten as

$$R_X(\tau) = \frac{\sigma_a^2}{T_s} \mathcal{R}_p(\tau) + \frac{\mu_a^2}{T_s^2} \sum_{-\infty}^{\infty} \mathcal{R}_p(\tau) * \delta(\tau - kT_s) \tag{4.10.14}$$

and take the Fourier transform to get

$$S_X(\omega) = \frac{\sigma_a^2}{T_s}|P(\omega)|^2 + \frac{2\pi \mu_a^2}{T_s^2} \sum_{k=-\infty}^{\infty} \left|P\left(\frac{2k\pi}{T_s}\right)\right|^2 \delta\left(\omega - \frac{2k\pi}{T_s}\right). \qquad (4.10.15)$$

Equations (4.10.13) and (4.10.15) find application in several problems in communications.

Summary

In this chapter we have briefly surveyed the topics of probability, random variables, and stochastic processes. This material is the absolute minimum knowledge required of these fields to continue our study of communication systems.

Problems

4.1 A binary data source S produces 0's and 1's independently with probabilities $P(0) = 0.2$ and $P(1) = 0.8$. These binary data are then transmitted over a noisy channel that reproduces a 0 at the output for a 0 in with probability 0.9, that is, $P(0|0) = 0.9$. The channel erroneously produces a 0 at its output for a 1 in with probability 0.2, that is, $P(0|1) = 0.2$.
   (a) Find $P(1|0)$ and $P(1|1)$.
   (b) Find the probability of a 0 being produced at the channel output.
   (c) Repeat part (b) for a 1.
   (d) If a 1 is produced at the channel output, what is the probability that a 0 was sent?

4.2 Rework Example 4.2.1 if $P(\xi_1) = 0.1$, $P(\xi_2) = 0.3$, $P(\xi_3) = 0.5$, and $P(\xi_4) = 0.1$.

4.3 For Example 4.2.1, define the random variables $X(\xi_1) = 0$, $X(\xi_2) = 1$, $X(\xi_3) = 2$, and $X(\xi_4) = 3$. Find and sketch the cumulative distribution function and the probability density function for this random variable.

4.4 Suppose that for Example 4.2.1, we are only interested in the number of telephones in use at any time. Hence we are led to define the random variable $X(\xi_1) = 0$, $X(\xi_2) = 1$, $X(\xi_3) = 1$, and $X(\xi_4) = 2$. Write and sketch the cumulative distribution function and the probability density function for this random variable.

4.5 We consider the random experiment of observing whether three telephones in an office are busy or not busy. The outcomes of this random experiment are thus:
   $\xi_1 = $ no telephones are busy.
   $\xi_2 = $ only telephone 1 is busy.
   $\xi_3 = $ only telephone 2 is busy.
   $\xi_4 = $ only telephone 3 is busy.

ξ_5 = telephones 1 and 2 are busy but 3 is not busy.

ξ_6 = telephones 1 and 3 are busy but 2 is not busy.

ξ_7 = telephones 2 and 3 are busy but 1 is not busy.

ξ_8 = all three telephones are busy.

The probabilities of these outcomes are given to be $P(\zeta_1) = 0.3$, $P(\zeta_2) = P(\zeta_3) = P(\zeta_4) = 0.1$, $P(\zeta_5) = P(\zeta_6) = P(\zeta_7) = 0.02$ and $P(\zeta_8) = 0.34$.

(a) What is the probability of the event that one or more telephones are busy?

(b) What is the probability that telephone 3 is in use?

(c) Consider the event that telephone 3 is busy and the event that only telephones 1 and 2 are busy. Are these two events statistically independent?

4.6 Suppose that for the random experiment in Problem 4.5, we are only interested in the number of telephones in use at any time. Define an appropriate random variable, find its cumulative distribution function and probability density function, and sketch both functions.

4.7 Given the distribution function of the discrete random variable Z,

$$
F_Z(z) = \begin{cases}
0, & z < 0 \\
\frac{1}{4}, & 0 \le z < 2 \\
\frac{3}{8}, & 2 \le z < 3 \\
\frac{1}{2}, & 3 \le z < 5 \\
\frac{7}{8}, & 5 \le z < 8 \\
1, & z \ge 8.
\end{cases}
$$

(a) Sketch $F_Z(z)$.

(b) Write an expression for $F_Z(z)$.

(c) Write an expression for and sketch the pdf of Z.

(d) Find $P[Z < 2]$, $P[Z \le 2]$, and $P[2 < Z \le 8]$.

4.8 The cumulative distribution function of a continuous random variable Y is sketched in Fig. PA.8.

(a) Write an expression for $F_Y(y)$.

(b) Find an expression for and sketch the pdf of Y.

(c) Find $P[Y \le 1]$, $P[Y \le 3]$, and $P[2 < Y \le 5]$.

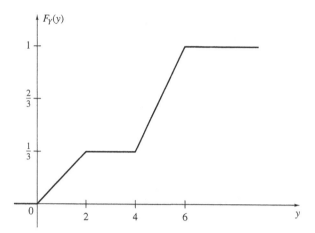

Fig. PA.8 Cumulative distribution function

4.9 Given the joint probability density function

$$f_{XY}(x, y) = \begin{cases} Ke^{-(3x+2y)/6}, & x \geq 0, \ y \geq 0 \\ 0, & x < 0, \ y < 0. \end{cases}$$

(a) Find K such that $f_{XY}(x, y)$ is a valid pdf.
(b) Find an expression for $F_{XY}(x, y)$.
(c) Find the marginal distribution and density functions of X.
(d) Calculate $P[X \leq 1, Y \leq 1]$.
(e) Calculate $P[X \leq 2, Y > 1]$.

4.10 Are the random variables X and Y in Problem 4.9 statistically independent? Use both distribution and density functions to substantiate your answer.

4.11 Given the joint pdf of Y and Z.

$$f_{YZ}(y, z) = \begin{cases} \frac{3y}{2} + \frac{z}{2}, & 0 \leq y \leq 1, \quad 0 \leq z \leq 1 \\ 0, & \text{otherwise.} \end{cases}$$

(a) Find the marginal densities of Y and Z.
(b) Find $f_{Y|Z}(y|Z)$.
(c) Calculate $F_{Y|Z}(y|Z = z)$.
(d) Calculate $F_{Y|Z}(y|z)$.
(e) Compare the results of parts (c) and (d).

4.12 Given the random variable X with pdf (α a Constant)

$$f_X(x) = Ke^{-\alpha x}u(x).$$

(a) Find K in terms of α.
(b) Calculate $E[X]$, $E[X^2]$, and var(X).
(c) What is $E\{2X^2 + 3X + 1\}$?

4.13 Compute the characteristic function of the pdf in Problem 4.12 and calculate the moments using Eq. (4.8).

4.14 Given the joint pdf

$$f_{XY}(x, y) = \begin{cases} \frac{1}{12}(2 + xy), & 0 \le x \le 2, \quad 0 \le y \le 2 \\ 0, & \text{otherwise.} \end{cases}$$

(a) Calculate $E\{XY\}$.
(b) Find cov(X, Y) and ρ_{XY}.
(c) Are X and Y uncorrelated? Orthogonal?

4.15 For the joint pdf in Problem 4.11:
(a) Find $E[YZ]$,
(b) Find cov(Y, Z) and ρ_{XY}.
(c) Are Y and Z uncorrelated? Orthogonal?

4.16 Given the discrete random variable Z in Problem 4.7, find the cumulative distribution function and the probability density function of the random variable $Y = Z + 2$.

4.17 The continuous random variable Y has the pdf

$$f_Y(y) = \frac{1}{2\sqrt{e}} e^{-(y-1)/2} u(y).$$

Find the pdf of the random variable $Z = 2Y + 2$.

4.18 Given the joint pdf of X and Y,

$$f_{XY}(x, y) = \begin{cases} 1, & 0 < x < 1, \quad 0 < y < 1 \\ 0, & \text{otherwise,} \end{cases}$$

Find the joint pdf of the random variables $W = X + Y$ and $Z = X - 7$.

4.19 Two random variables X and Y have the joint pdf

$$f_{XY}(x, y) = \begin{cases} \frac{1}{4} e^{-(x+y)/2}, & 0 \le x < \infty, \quad 0 \le y < \infty \\ 0, & \text{otherwise.} \end{cases}$$

We wish to find the pdf of $W = X + Y$.
     Hint: Use an auxiliary variable.

4.20 A double-sided exponential or Laplacian random variable has the pdf

$$f_X(x) = K_1 e^{-K_2|x|}, \qquad -\infty < x < \infty.$$

(a) For this to be a pdf, how are K_1 and K_2 related? Let $K_2 = 1$ and find K_1.
(b) Is X zero mean?
(c) Compute the variance of X.
4.21 Find the characteristic function of the Laplacian random variable

$$f_Y(y) = \frac{1}{2\sigma} e^{-|y-\mu|/\sigma}, \qquad -\infty < y < \infty,$$

where $\sigma > 0$. Calculate the mean and variance.
4.22 A Gaussian random variable X has the pdf

$$f_X(x) = \frac{1}{\sqrt{2\pi}\sigma} e^{-(x-\mu)^2/2\sigma^2}, \qquad -\infty < x < \infty.$$

Find the pdf of the random variable $Y = aX + b$.
4.23 A random variable X has the uniform pdf given by

$$f_X(x) = \begin{cases} \frac{1}{b-a}, & a \leq x \leq b \\ 0, & \text{otherwise.} \end{cases}$$

(a) Find the pdf of $X_1 + X_2$ if X_1 and X_2 are independent and have the same pdf as X.
(b) Find the pdf of $X_1 + X_2 + X_3$ if the X_i, $i = 1, 2, 3$ are independent and have the same pdf as X.
(c) Plot the resulting pdfs from parts (a) and (b). Are they uniform?
4.24 Let $Y = \sum_{i=1}^{P} X_i$, where the X_i are independent, identically distributed Gaussian random variables with mean 1 and variance 2. Write the pdf of Y.
4.25 Derive the mean and variance of the Rayleigh distribution in Eqs. (4.6.17) and (4.6.18).
4.26 Find the mean and variance of the binomial distribution using the characteristic function in Eq. (4.6.24).
4.27 For a binomial distribution with $n = 10$ and $p = 0.2$, calculate, for the binomial random variable X, $P[X \leq 5]$ and $P[X < 5]$. Compare the results.
4.28 Use the characteristic function of a Poisson random variable in Eq. (4.6.31) to derive its mean and variance.
4.29 A random process is given by $X(t) = A_c \cos[\omega_c t + \Theta]$, where A_c and ω_c are known constants and Θ is a random variable that is uniformly distributed over $[0, 2\pi]$.
(a) Calculate $E[X(t)]$ and $R_X(t_1, t_2)$.
(b) Is $X(t)$ WSS?
4.30 For the random process $Y(t) = Ae^{-t}u(t)$, where A is a Gaussian random variable with mean 2 and variance 2,
(a) Calculate $E[Y(t)]$, $R_Y(t_1, t_2)$, and $\text{var}[Y(t)]$.
(b) Is $Y(t)$ stationary? Is $Y(t)$ Gaussian?

4.31 Let X and Y be independent, Gaussian random variables with means μ_x and μ_y and variances σ_x^2 and σ_y^2, respectively. Define the random process $Z(t) = X \cos \omega_c t + Y \sin \omega_c t$, where ω_c is a constant.

(a) Under what conditions is $Z(t)$ WSS?

(b) Find $f_z(z; t)$

(c) Is the Gaussian assumption required for part (a)?

4.32 A weighted difference process is defined as $Y(t) = X(t) - \alpha X(t - \Delta)$, where α and Δ are known constants and $X(t)$ is a WSS process. Find $R_Y(\tau)$ in terms of $R_X(\tau)$.

4.33 Use the properties stated in Sect. 4.7 to determine which of the following are possible autocorrelation functions for a WSS process ($A > 0$).

(a) $R_X(\tau) = Ae^{-\tau}u(\tau)$

(b) $R_X(\tau) = \begin{cases} A[1 - |\tau|/T], & |\tau| \le T \\ 0, & \text{otherwise} \end{cases}$

(c) $R_X(\tau) = Ae^{-|\tau|}, \quad -\infty < \tau < \infty$

(d) $R_X(\tau) = Ae^{-|\tau|} \cos \omega_c \tau, \quad -\infty < \tau < \infty$

4.34 It is often stated as a property of a WSS, nonperiodic process $X(t)$ that

$$\lim_{\tau \to \infty} R_X(\tau) = (E[X(t)])^2.$$

This follows, since as τ gets large, $X(t)$, $X(t + \tau)$ become uncorrelated. Use this property and property 1 in Sect. 4.7 to find the mean and variance of those processes corresponding to valid autocorrelations in Problem 4.33.

4.35 Calculate the power spectral densities for the admissible autocorrelation functions in Problem 4.33.

4.36 Calculate the power spectral density for Problem 4.31(a).

4.37 If $\mathcal{F}\{R_X(\tau)\} = S_X(\omega)$, calculate the power spectral density of $Y(t)$ in Problem 4.32 in terms of $S_X(\omega)$.

4.38 A WSS process $X(t)$ with $R_x(\tau) = V\delta(\tau)$ is applied to the input of a 100% roll-off raised cosine filter with cutoff frequency ω_{co}. Find the power spectral density and autocorrelation function of the output process $Y(t)$.

4.39 A WSS signal $S(t)$ is contaminated by an additive, zero-mean, independent random process $N(t)$. It is desired to pass this noisy process $S(t) + N(t)$ through a filter with transfer function $H(\omega)$ to obtain $S(t)$ at the output. Find an expression for $H(\omega)$ in terms of the spectral densities involving the input and output processes.

References

1. Gray, R. M., and L. D. Davisson. 1986. *Random Processes: A Mathematical Approach for Engineers.* Englewood Cliffs, N.J.: Prentice Hall.
2. Franks, L. E. 1969. *Signal Theory.* Englewood Cliffs, N. J.: Prentice Hall.

Printed in the United States
by Baker & Taylor Publisher Services